The Complete Guide to the New NCO-ER

By Wilson L. Walker

*Becoming a Better Leader and Getting Promoted
in Today's Army*

The Complete Guide to the New NCO-ER

The NCO-ER Leadership Guide

The Self-Development Test Study Guide

*Up or Out: How to Get Promoted as
the Army Draws Down*

The Complete Guide
to the New NCO-ER

How to Receive and Write an Excellent Report

Second Edition

Wilson L. Walker
Master Sergeant, U.S. Army, Retired

IMPACT PUBLICATIONS
Manassas Park, Virginia

Second Edition

Library of Congress Cataloguing-in-Publication Data

Walker, Wilson L., 1944-
   The complete guide to the new NCO-ER: how to receive and
write an excellent report / Wilson L. Walker
        p. cm.
     ISBN 1-57023-202-4
     1. United States. Army – Non-commissioned officers – Rating
of – management – Handbooks, manuals, etc. I. Title.
   UB323.W277
   355.3'38 – dc20   2002109080

Publisher: For information on Impact Publications, including current and forthcoming publications, authors, press kits, online bookstore, and submission requirements, visit our website: www.impactpublications.com.

Publicity/Rights: For information on publicity, author interviews, and subsidiary rights, contact the Media Relations Department: Tel. 703-361-7300, Fax 703-335-9486, or email: info@impactpublications.com.

Sales/Distribution: All sales and distribution inquiries should be directed to the publisher: Sales Department, IMPACT PUBLICATIONS, 9104 Manassas Drive, Suite N, Manassas Park, VA 20111-5211, Tel. 703-361-7300, Fax 703-335-9486, or email: info@impactpublications.com.

Contents

Acknowledgments

A S ALWAYS, I THANK GOD who makes it all possible for all of us. I also would like to thank my mother, wife, and children.

I have a very special thanks for Tammie whose editing enabled her to translate what I try to say into words that all soldiers can understand and receive the information intended. Also, I would like to thank Mardie Younglof and Ron Krannich at Impact Publications for editing the book as well as for turning the information into a book and placing it where it is available to all soldiers.

This acknowledgment would not be complete without thanking all the soldiers for their support in ensuring we all live in a peaceful world. You are the best, and we all wish you success and happiness in the years ahead.

Wilson L. Walker
Master Sergeant, U.S. Army, Retired

The Complete Guide to the New NCO-ER

1

The NCO-ER

THE NON-COMMISSIONED OFFICER Evaluation Report (NCO-ER) is one – if not **the** – most important reports that you will receive or write as an NCO in the Army. The Noncommissioned Officer Evaluation Reporting System is designed to:

1. Strengthen the ability of the NCO corps to meet the professional challenges of the future through the indoctrination of Army values and basic NCO responsibilities.

2. Ensure the selection of the best qualified noncommissioned officers to serve in positions of increasing responsibility by providing the rating officials a view of performance/potential for use in centralized selection, assignments, and other Enlisted Personnel Management System decisions.

3. Contribute to Army-wide improved performance and professional development by increased emphasis on performance counseling.

4. Recognize that continuous professional development and
    growth best serve the Army and the NCO.

Evaluation reports must be accurate and complete to ensure that
sound personnel management decisions can be made and that an
NCO's potential can be fully developed. Each report must be at
thoughtful, fair appraisal of an NCO's ability and potential.

The following chapters outline how you can write and receive an
excellent NCO-ER for getting promoted and receiving good assignments in today's Army. Each chapter outlines important principles and
procedures you should follow for advancing your career in the Army.

2

The Rating Officials

THE RATING OFFICIALS WILL consist of the rater, the senior rater, and the reviewer. Their mission is to provide the best evaluation of an NCO's performance and potential. The rating officials also tie the rated NCO's performance to a specific senior/subordinate relationship. Rating is best achieved within an organization's chain of command. Individuals in non-supervisory positions are not authorized to evaluate NCOs.

The Rater

What exactly does the rater do in reference to the NCO? The rater must:

- Be the immediate supervisor of the rated NCO.

- Be designated as the rater for a minimum period of 90 rated days.

- Be a sergeant or above and senior to the rated NCO by pay grade or date of rank.

- Be familiar with the day-to-day performance of the rated NCO, and must directly guide the rated NCO's participation in the organization mission.

- Prepare complete, accurate, and fully considered evaluation reports.

- Make an honest, fair evaluation of the NCO under his/her evaluation supervision.

- Be obligated to the NCO and the Army and be honest in his/her evaluation.

- Prepare a fair, correct report evaluating the NCO's duty performance, value/NCO responsibilities, and potential.

- Verify part I and II and enter the PT and height and weight results in part IVc of the NCO-ER.

- Sign part IIb and d when serving as the senior rater and reviewer.

The primary role of the rater is that of evaluation, focusing on performance and performance counseling. The rater will counsel the rated NCO on his/her duty performance and professional development throughout the rating period. The rater also will define and discuss the duty description for part III of the NCO-ER with the rated NCO during these sessions. At a minimum, the rated NCO will be counseled within the first 30 days of each rating period and quarterly thereafter. Corporals and sergeants will be counseled within the first 30 days of the effective date of lateral appointment to corporal or promotion to sergeant. The rater will prepare a separate DA Form 2166-8-1 for each rated NCO, and use the form together with the working copy of the NCO-ER for conducting performance counseling. The DA Form 2166-8-1 is mandatory for use by the rater when counseling NCOs, Cpl through CSM. CSMs will not be rated if they work for a three- or four-star commander.

The DA Form 2166-8-1 will be maintained by the rater until the NCO-ER for that period has been approved and submitted to USAEREC at DA. For corporals, who will not receive a record NCO-ER, the rater will maintain the checklist for one year, and in some cases keep it for possible future use to support personnel action may be appropriate. If the rater is on a recommended list for promotion to one of the top three NCO grades and is serving in an authorized position for the new grade, then he/she may rate any NCO he or she supervises, if, after the rater's promotion, he/she will be senior in pay grade or date of rank to the rated NCO. An NCO frocked to the grade of ISG, SGM, or CSM and is serving in an authorized ISG, SGM, or CSM position may rate any NCO he/she supervises, if after promotion he/she will be senior to the rated NCO by either pay grade of date or rank. Commanders may appoint civilian employees of DOD, GS-6 and above, as raters when an immediate military supervisor is not available and when the civilian supervisor is in the best position to accurately evaluate the NCO's performance. The civilian rater must be officially designated on the published rating scheme established by the local commander.

The Senior Rater

The senior rater uses his/her position and experience to evaluate the rated NCO from a broad organizational perspective. His/her evaluation is a link between the day-to-day observation of the rated NCO's performance by the rater and the long-term evaluation of the rated NCO's potential by the DA selection board. Normally, to evaluate an NCO, the senior rater must be designated and serve in the capacity for at least 60 rated days. The senior rater's primary role is to evaluate potential, over-watch the performance evaluation, and mentor subordinates. The senior rater must:

- Use all reasonable means to become familiar with the rated NCO's performance throughout the rating period.

- Prepare a fair, correct report evaluating the NCO's duty performance, professionalism, and potential.

- Date and sign the report in part Iib.

- Obtain the rated NCO's signature in part II of the NCO-ER.

- Be in the direct line of supervision of the rated NCO and designed as the senior rater for a minimum period of 60 rated days.

- Ensure the specific bullet example support the appropriate rating in part Ivb-f.

- Ensure the bullet "senior rater does not meet minimum qualification" is entered in part Ve when the senior rater does not meet the minimum requirement.

- Not render an evaluation in part Vc or d when the minimum time requirement is not met.

- Sign part IId when also serving as reviewer.

- Not direct the rater tc change an evaluation that he/she believe to be honest.

- Ensure the rated NCO is aware that his/her signature does not constitute agreement or disagreement with the evaluation of the raters.

The rated NCO's signature ONLY indicates that he/she has seen the completed report (except parts IId-e), has verified that the administrative data (part I) is correct, the rating officials are proper (part II), and the duty description is accurate (part II!) and includes the counseling dates. When the counseling dates are omitted, the senior rater will enter a statement in part Ve, explaining why counseling was not accomplished. The NCO's signature also verifies that the APFT and height/weight entries are correct (part IVc) and shows awareness of the appeals process contained in Chapter 5. If the NCO refuses to sign the report or is unavailable to sign the report, enter the appropriate statement "NCO refuses to sign" or "NCO is not available for signature" in (part IIc). Commanders may appoint civilian employees of DOD, GS-6 and above, as senior raters when a military supervisor is not being evaluated and when the civilian supervisor is in the best position to accurately evaluate the NCO's performance. The civilian senior rater must be officially designated on the published rating scheme established by the local commander. Members of other U.S. military services who meet the qualifications may be senior raters. A rater may act as both the rater and senior rater, when the rater is a general officer, an officer of flag rank, or a civilian with Senior Executive Services (SES) rank and precedence. In most cases the senior rater will be the rater's rater, or the rater's boss.

The Reviewer

The reviewer must be a commissioned officer, warrant officer, command sergeant major, or sergeant major in the direct line of supervision and senior in pay grade or date of rank to the senior rater. Most of the time the reviewer is the senior rater, rater, or boss. A promotable master sergeant may serve as reviewer, provided they are working in an authorized CSM or SGM position. No minimum time period is required for the reviewer qualification. The command can also appoint officers of another military service or civilian employees of DOD, GS-6 and above or other civilian pay scales as determined by the commander, as reviewers when the grade and line of supervision requirement are met, and either the rater or senior rater is a uniformed army official. In cases where both the rater and senior rater are other than uniformed army rating officials, and no uniformed army reviewer is available, the report will be reviewed by a uniformed army officer in the rated NCO's PSB or unit administrative office; this officer is not required to be senior to the rater or senior rater. If the rater or senior rater is a general officer, he/she can also act as reviewer. The reviewer will ensure that the proper rater and senior rater complete the report. He/she will examine the evaluation rendered by the rater and senior rater to ensure that they are clear, consistent, and just, in accordance with known facts. Special care must be taken to ensure the specific bullet comments support the appropriate excellence, success, or need improvement rating in part IVb-f. The reviewer will indicate concurrence or nonconcurrence with the rater and/or senior rater by annotating the appropriate box with a typed or handwritten "X" in part II and adding an enclosure, when the nonconcurrence box is marked. If the reviewer determines that the rater and senior rater have not evaluated the rated NCO in a clear, consistent, or just manner based on known facts, the reviewer's first responsibility is to consult with one or both rating officials to determine the basis for the apparent discrepancy. If the rater and/or senior rater acknowledge the discrepancy and revise the NCO-ER so that the reviewer agrees with the evaluation, then the reviewer will check the concur box in part II. If the rater and/or senior rater fail to acknowledge a discrepancy and indicate that the evaluation is their honest opinion, the reviewer checks the nonconcur box in part II. The reviewer then add an enclosure that clarifies the situation and renders his/her opinion regarding the rated NCO's performance and potential. The reviewer may not direct that the rater and/or senior rater change an evaluation believed to be honest. In cases where neither rater nor senior rater is an NCO, the reviewer may find it useful to get additional informal input from the senior NCO subordinate to the reviewer.

Loss of a Rating Chain Official

Special rules apply when a rating chain official is unable to render an evaluation on the rated NCO. These situations occur when a rating official:

- dies

- is suspended

- relieved

- absent without leave (AWOL)

- declared missing

- declared incapacitated

When a rating official is relieved, reduced, AWOL, or incapacitated, he/she will not be permitted to evaluate his/her subordinates. When the senior rater or reviewer is removed from the rating chain, a new rating official is designated, and may participate after the minimum required time is met. When the rater or senior rater is suspended, the suspended time will be counted as non-rated time. When the rater is removed from the rating chain, and the minimum time has not been met, the period is non-rated and a new rater is designated. If the minimum time is met, the senior rater can perform the functions of the rater and the senior rater. If the senior rater serves as the rater, the rating period of the report will be the period the senior rater has been in the rating chain.

3

DA Form 2166-8 and DA Form 2166-8-1

NCO Counseling Checklist/Record

The purpose of the checklist is to improve performance counseling by providing structure and discipline as well as improving performance. The rater uses DA Form 2166-8-1, along with a working copy of the NCO-ER (DA Form 2166-8), to prepare for, conduct, and record results of performance counseling with the rated NCO. If proper counseling is being done, a rated NCO should not be alarmed if he/she receives a report that highlights shortcomings or failures. The best counseling does not dwell on the past and what "was" done, but rather on the future and what "can" be done. The NCO counseling checklist/record contains key and essential information to prepare for and conduct a counseling session. The process for the rated NCO counseling is as follows:

1. Within the first 30 days of the rating period, effective date of lateral appointment to corporal, or promotion to sergeant, the rater will conduct the first counseling session

with the rated NCO. The rater should specifically let the rated NCO know what is expected during the rating period. The DA Form 2166-8-1 provides examples, definitions, and step-by-step assistance to the rater for preparing and communicating performance standards and direction to the rated NCO. The rater should show the rated NCO the rating chain and a complete duty description, discuss the meaning of the values (see Army Values in Chapter 4) and responsibilities contained on the NCO-ER, and explains the standards for success. Before the rated NCO departs the counseling session, the rater records key points discussed and obtains the rated NCO's initials on page 2 of the DA Form 2166-8-1.

2. The rater will conduct later counseling sessions during the rating period; they will be conducted at least quarterly. These counseling sessions differ from the first counseling session in that the primary focus is on telling the rated NCO how well he/she is doing. The DA Form 2166-8-1 provides step-by-step assistance to the rater. The rater updates the duty description and, based on observed action and demonstrated behavior and results, discusses what was done well and what could be done better. The guide for this discussion is the success standards established in the previous counseling session the rater records key points discussed and obtains the rated NCO's initials on page 2 of the DA Form 2166-8-1.

3. The rater will maintain one DA Form 2166-8-1 for each NCO until after the NCO-ER for that period has been approved and submitted to USAEREC. For the corporals, who do not receive a record NCO-ER, the checklist will be maintained for one year.

Rater and Rated NCO Counseling

In order to improve or maintain performance and professionally develop the rated NCO, face-to-face performance counseling between the rater and the rated NCO must be accomplished. The rater develops and communicates performance standards to the rated NCO at the beginning of the rating period. The rater should also conduct additional performance counseling during the rating period by providing the rated NCO with feedback regarding his/her progress in meeting the goals established at the beginning of the rating period. The goals of performance counseling is to get all NCOs to be successful and meet or exceed standards. Face-to-face performance counseling is mandatory for all NCOs. The initial counseling will be accomplished within the first 30 days of the rating period, and additional counseling will be conducted at least quarterly (every three months) thereafter. The rater should also include the commander's special interest items during the counseling session, if there are any.

Ending the Counseling Session

Before the NCO departs the counseling session, the rater should:

- Record the counseling date on DA Form 2166-8.

- Write any additional key points that came up during the counseling session.

- Show key points to the rated NCO and get his/her initials.

- Save the NCO-ER with the checklist for the next counseling session.

Preparing for the First Counseling Session

Before the rating NCO starts counseling the rated NCO, there are things that have to be done, such as:

1. Schedule the counseling session, and notify the rated NCO.

2. Get a copy of the last duty description used for the rated NCO's duty position, a blank copy of the NCO-ER, and the names of the new rating chain.

3. Update the duty description if needed.

4. Fill out the rating chain and duty description on the working copy of the NCO-ER, parts II and III.

5. Read each of the values/responsibilities in part IV of the NCO-ER and the expanded definitions and examples on page 3 and 4 of the NCO counseling checklist (DA Form 2166-8-1).

6. Think how each value and responsibility in part IV of the working copy of the NCO-ER applies to the rated NCO and his/her duty position.

7. Decide what is considered necessary for success rating for each value and responsibility.

8. Make notes in blank spaces in part IV of the working copy of the NCO-ER to help when counseling the rated NCO.

9. Record the rated NCO's name, rank, duty position, and unit data on page one of DA Form 2166-8-1.

10. Write key points to be made during the counseling session on DA Form 2166-8-1, and review development counseling (see Chapter 5).

During the First Counseling Session

1. Make sure the rated NCO knows the rating chain.

2. Show the rated NCO the draft duty description on the working copy; explain all parts, if the rated NCO performed in that position before; ask for any ideas to make the duty description better.

3. Discuss the meaning of each value/responsibility in part IV of the working copy. Use the trigger words on the NCO-ER

and the expanded definitions on pages 3 and 4 of DA Form 2166-8-1.

4. Explain how each value/responsibility applies to the specific duty position by showing or telling your standards for success. Be specific so the NCO really knows what's expected of him/her.

5. When possible, give specific examples of excellence that could apply. "Remember that only a few achieve real excellence."

6. Give the rated NCO an opportunity to ask questions.

Before the NCO Departs the Counseling Session

1. Record the counseling date on DA Form 2166-8-1.

2. Write any additional key points that came up during the counseling session.

3. Show key points to the rated NCO and get his/her initials.

4. Save DA Form 2166-8 and DA Form 2166-8-1 for the next counseling session.

Preparing for Later Counseling Sessions

1. Schedule the counseling session, notify the rated NCO, and tell him/her to come prepared to discuss what has been accomplished in each value area.

2. Look at the working copy of the NCO-ER you used during the last counseling.

3. Read and update the duty description, especially noting the area of special emphasis.

4. Read again each value/responsibility in part IV of the working copy. See the definitions and examples on pages 3 and 4 of DA Form 2166-8-1.

5. Look over notes you wrote down on page 2 of DA Form 2166-8-1 about the last counseling session.

6. Think about what the rated NCO has done so far during the rating period (specifically, observed action, demonstrated behavior, and results).

For each value/responsibility area, answer three questions:

1. What has happened in response to any discussion you had during the last counseling session?

2. What has been done well?

3. What could be done better?

Make notes in the blank spaces in part IV to help focus when counseling.

1. Write the key point to be made during the counseling session.

2. Review development counseling in Chapter 4.

During the Later Counseling Session

1. Go over each part of the duty description with the rated NCO. Discuss any changes, especially in the area of special emphasis.

2. Tell the rated NCO how he/she is doing. Use your success standards as a guide for the discussion, first for each value/responsibility, talk about what has happened in response to any discussion you had during the last counseling session. Next, talk about what was done well, and how

things can be done better. Remember the Army's goal is to get all NCOs to be successful and meet standards.

3. When possible, give examples of excellence that could apply. This gives the rated NCO something to strive for, but, remember, "excellence" is very special. Only a few achieve it. Excellence includes results and often involves subordinates.

4. Ask the rated NCO for ideas, examples, and opinions on what has been done so far, and what can be done better.

Before the Rated NCO Departs the Counseling Session

1. Record the counseling date on DA Form 2166-8-1.

2. Write any additional key points that came up during the counseling session.

3. Show the key points to the rated NCO and get his/her initials on DA form 2166-8-1.

4. Save DA Forms 2166-8-1 and 2166-8 for the next counseling session.

Working Copy of the NCO-ER

The working copy of the NCO-ER is nothing more than *DA Form 2166-8-1* or the *NCO Evaluation Report*. It's called the working copy because it is used along with DA Form 2166-8-1 when counseling an NCO or corporal. When the rater starts out with DA Form 2166-8, the first thing he/she will do is look over Part One.

Part One

ADMINISTRATIVE DATA

The administrative data includes identifying the rated NCO, the period of the report, and the reason for submitting the report. The rater will verify the data in Part I with the rated NCO and notify the Battalion S-1 or administrative office of any errors. Now let's look over the blocks. We will fill them all in so that not only will you know which blocks to fill in for the working copy but also those used for your NCO-ER.

1. **Parts Ia and Ib are self-explanatory.** The name will be capitalized.

2. **Part Ic.** Here you put in the three-letter abbreviation for the NCO's rank (example: SGT, SSG, MSG). If the rated NCO is frocked to ISG, SGM, or CSM, enter the rank, date of rank, and PMOSC held prior to the frocking action. Also in addition to the NCO's rank in Part Ic, enter the appropriate rank in parentheses next to the rank entry (example: SFC [ISG] MSG [SGM]).

3. **Part Id.** Use six digits to enter the rated NCO's date of rank (example: 030115). If the rated NCO is frocked, enter the date of rank for the rank held prior to the frocking action.

4. **Part Ie.** Enter up to nine digits of the primary MOS code if the rated NCO possesses an additional skill identifier; if not, only a five-digit MOS code is needed (example: 75H5MA3 or 24T40).

5. **Part If.** Here you put in the rated NCO's military address in the order listed in block If: Unit, Org, Station, Zip Code, or APO, Major Command.

6. **Part Ig.** Put the appropriate NCO-ER code in the left side of the block and the report title on the right side portion of

the block example: 03 Change of Rater).

7. **Part Ih.** (From date) The beginning month is the month following the ending month of the last report. Use a four-digit numerical identifier for the year (2003), and a two-digit numerical identifier for the month (09), so September 2003 will become 2003 09. (Thru-date) should be entered in the same manner as the from date. The ending month is the month of the event generating the report, regardless of when the event occurs during the month.

8. **Part Ii.** Compute the number of rated months as shown below.

Non-Rated Days	**to**	**Non-Rated Months**
15 days or less		0 months
16-45 days		1 months
46-75 days		2 months
76-105 days		3 months
106-135 days		4 months
136-165 days		5 months

Subtract the non-rated months from the total months; the remainder is the number of rated months.

9. **Part Ij.** Enter the appropriate codes. If there were no non-rated periods, leave Ij blank. Non-rated codes and reasons are as follows:

Non-Rated Code	**Reason for Code**
Code A	AWOL/desertion/unsatisfactory partici-pant based on AR 135-91.
Code B	Break in active enlisted service or 12 months or less.

Code C	Confinement in detention facility; assignment personnel control or correctional training facility.
Code D	Temporary disability retirement list (TDRL) status.
Code I	In transit between duty stations, including leave and TDY.
Code M	Missing in action.
Code P	Patient (including convalescent leave).
Code Q	Lack of rater qualification.
Code R	New Recruiter Program (see AR 601-1).
Code S	Student at military or civilian school.
Code W	Prisoner of war.
Code X	Inactive National Guard, Reserve, or Standby Reserve.
Code Z	Other approved reasons. This code can alsobe used when there is a non-rated period of less than 12 months resulting from a reduction to a rank below SGT or when a previous command did not render an NCO-ER.

10. **Part Ik.** If there are any authorized enclosures to be attached and forwarded with the NCO-ER, put the number in this block (Ik).

11. **Part Ii.** Enter the handwritten or typewritten "X" in block "Given to NCO," and the six-digit date of the same

block (2) if the rated NCO copy is to be forwarded.

12. **Part Im.** Enter the handwritten PSB/RC representative's initials.

13. **Part In.** Enter the rated NCO's major command by entering the two-character command assignment code (see AR 680-29).

14. **Part Io.** Enter the four-position alphanumeric PSC code.

Part Two

AUTHENTICATION

Part Two is for authentication by the rated NCO and rating officials after they have completed their portions of the NCO-ER at the end of the rating period. Restrictions on signature dates are discussed in Chapter 4 (type of reports). Rated NCOs and rating officials should not sign blank NCO-ER forms. Instructions for Part II "Authentication" are as follows:

1. **Part Iia, b, and d** is self-explanatory. (See #5.)

2. **Part IIc.** Here the rater will verify Parts I and II, and the APFT and the height and weight entries with the rated NCO. The senior rater will obtain the rated NCO's signature or enter the appropriate statement "NCO refuses to sign" or "unavailable for signature."

3. **Part IId** also is self-explanatory.

4. **Part IIe.** Here the reviewer will place a typed or handwritten "X" in the appropriate block, indicating concurrence. The reviewer will also ensure the rated NCO is provided a copy of the non-concurrence enclosure, if need be.

5. **Part Iia, b, and d.** The rank portion of Part IIa, b, and d will contain the appropriate three-letter Army abbreviation unless the official is a promotable master sergeant occupying a sergeant major position and acting as a reviewer, in which case enter MSG (P).

Part Three

DUTY DESCRIPTION

Part III provides for the duty description of the rated NCO. It is the responsibility of the rating officials to ensure the duty description information is factually correct.

The duty description:

1. Is entered by the rater and verified with the rated NCO.

2. Is an outline of the normal requirements of the specific duty position.

3. Should show type of work required rather than frequently changing tasks.

4. Is essential to performance counseling and evaluation.

5. May be updated during the rating period.

6. Is used at the end of the rating period to record what was important about the duties.

Now let's go over Part III, just like we did for Parts I and II. We'll fill in the blanks as we go. The detailed instruction are as follows:

1. **Part IIIa.** Enter the principal duty title that most accurately reflects actual duties performed.

2. **Part IIIb.** Here is the block for the DMOS. It should be entered using five to nine characters.

3. **Part IIIc.** Daily duties and scope must be a series of phrases, starting with action words and separated by semicolons. The most important routine duties and responsibilities should be addressed. The number of soldiers supervised should be included in Part IIIc, as well as equipment, facilities, and dollars involved and any other routine duties and responsibilities critical to the mission accomplishment.

4. **Part IIId.** An area of special emphasis is the block for any appointed duties or tasks the rated NCO may have. It should also include the most important items that applied at any time during the rating period, such as NTC rotation, FTX preparation, AGI, and combined arms drills, and such.

5. **Part IIIe.** Appointed duties are those that are appointed and are not normally associated with the duty description such as Key control, Safety NCO, and Training NCO.

6. **Part IIIf.** The counseling dates are the actual dates from DA Form 2166-8-1. There is a block for initial and later counseling.

Part Four

ARMY VALUES/NCO RESPONSIBILITIES

Part IV is completed by the rater, including the APFT performance entry and the height and weight entry in Part IVc. Part IVa contains a listing of the Army Values that define professionalism for the Army NCOs. They will be considered in the evaluation of the performance of all NCOs. Values and NCO responsibilities/requirements are the sole focus for evaluation of performance in Part IV of the NCO-ER. The detailed instructions for Part IV are as follows:

1. **Part Iva.** The rater will check either a "yes" or "no" in the value blocks. Mandatory specfic bullet comments are required for all "no" entries. Every entry should be based on whether the rated NCO "meets" or "does not meet" the standards for each particular value. The list of the Army Values and their definition can be found in Chapter five under the title "Army Values."

2. **Part Iv b, c, d, e, and f.** In these blocks the rater will indicate the level of performance for each responsibility by placing a type and/or handwritten "X" in the appropriate box. Performance levels are:

   A. **Excellence.** Exceeds standards; demonstrated by specific examples and measurable results; special and unusual; achieved by only a few NCOs, clearly better then most others.

   B. **Success.** Meets all standards. Majority of ratings are in this category; fully competitive for schooling and promotion. The main goal of counseling is to bring all NCOs to this level.

   C. **Need improvement.** Missed some standards. This is one rating you don't need on your report. If you do, make sure the next five in that area are successful.

Part Four "C"

ARMY PHYSICAL FITNESS TEST ENTRY

1. The rater will enter one of the following APFT entries, "PASS" or "FAIL," and year and month of the APFT results. APFT refers to both the PT test for NCOs without profiles consisting of push-ups, sit-ups, and the two-mile run; and the alternate PT test as prescribed by health care personnel for NCOs with permanent profiles who have been cleared to take the alternate PT test. If no APFT is

taken due to profiles, the entry will be "PROFILE" and the year and month of the profile or when the profile was awarded. These entries will reflect the NCO's status on the date of the most recently recorded APFT administered by the unit within the 12-month period prior to the last rated day of supervision. Sample entries are "PASS 0105, FAIL 0105," or "PROFILE 9904." NCOs who have a permanent profile and are cleared to take the alternate PT test do not need the statement "Profile does or does not hinder duty performance." The APFT is considered valid as long as it consists of one of the aerobic events (run, walk, bike, or swim). The APFT badge is awarded for scores of 270 and above, with at least a score of 90 in each event. "Received APFT badge" may be entered as a bullet comment to justify an "excellence" rating. Numerical scores will be used to justify a "need improvement" rating that are based solely on the APFT. The scores are optional for the "success" rating.

2. Rater specific bullet examples are mandatory in Part IVc for the following:

A. APFT entry of "FAIL" or "PROFILE."

B. The absence of an APFT entry if the APFT has not been taken within 12 months of the "THRU" date of the report.

C. Exempt from APFT requirement if no APFT has been taken within the last 12 months due to pregnancy, convalescent leave, and temporary profile.

3. Personnel who meet the army minimum standards for APFT, but fail to meet the unit standards, will not be given a rating of "need improvement" for physical fitness and military bearing if such a rating is based solely on the failure to meet UNIT standards.

HEIGHT AND WEIGHT ENTRY

The rater will enter the rated NCO's height and weight as of the unit's last record weigh-in and an entry of "YES" or "NO" to indicate compliance or noncompliance IAW AR 600-9. If there is no record weight-in during the period covered by the report, the rater will enter the NCO's height and weight as of the "THRU" date of the NCO-ER. The rater will enter "YES" for those NCOs who meet the weight and height screening table or are in compliance with the body fat standards of AR 600-9.

The rater specific bullet examples are mandatory in IVc to:

1. Explain any entry of "NO," indicating noncompliance with the standards of AR 600-9. Medical waivers to weight control standards are not permitted for evaluation reports, so the "NO" entry is required. Rating officials will not use the word "pregnant" or refer to an NCO's pregnancy in any manner when completing an NCO-ER.

2. Explain the absence of the height and weight data.

Part Five

OVERALL PERFORMANCE AND POTENTIAL

The rater will place a "X" in the appropriate box. NCOs receiving one or more "Need Improvement" ratings in Part IVb-f cannot receive a rating of "among the best." The following definition will be used when completing Part Va.

1. **Among the best.** This rating is for NCOs who have demonstrated a good performance, have a strong recommendation for promotion should sufficient allocations be available.

2. **Fully capable.** This rating is for NCOs who have demonstrated a good performance, have a strong recommendation

for promotion should sufficient allocation be available.

3. **Marginal.** This rating is for NCOs who have demonstrated poor performance and should not be promoted at this time.

In Part Vc and Vd, the senior rater will evaluate the overall performance and potential by placing an "X" in the appropriate box for each area. The following definitions will be used when completing Part Vc and Vd:

1. **Successful/Superior.** A "1" rating represents the cream of the crop and is a recommendation for immediate promotion. A "2" rating represents a very good, solid performance and is a strong recommendation for promotion. A "3" rating represents a good performance, and should there be sufficient allocations available, can be promoted.

2. **Fair.** This rating represent NCOs who may require additional training, observation, and should not be promoted at this time.

3. **Poor.** This rating represents NCO's who are weak or deficient and, in the opinion of the senior rater, needs significant improvement or training in one or more areas.

4

Bullet Comments

A BULLET COMMENT IS A STATEMENT THAT MAY OR may not have a verb, object, or subject. It is a short, concise comment used by raters to justify their evaluations. Because they require the raters to make specific points, bullet comments are hard to inflate.

Bullet comments will:

1. Be short, concise, to the point. They will not be longer than two lines, preferably one, and no more than one bullet to a line.

2. Start with action words (verbs) or possessive pronouns (his/her); do not use "past" tense when addressing NCO's performance and/or contributions.

3. Be double-spaced between bullets.

4. Be preceded by a small "o" to designate the start of the comment. Each bullet comment should start with a small

letter unless it is a proper noun, which will usually be capitalized.

5. Before completing the evaluation portion of Part IV, the following must be considered:

- Values/NCO Responsibility
- Commander's Evaluation (CE)
- Common Task Test (CTT)
- Army Physical Fitness Test (APFT)
- Weapon Qualifications
- Army Weight Control Program (AR 600-9)

The rater must explain, with specific bullet examples, any area where the rated NCO demonstrated excellence, notable success, or needs improvement. Specific bullet examples can be used only once; therefore, the rater must decide under which responsibility the bullet fits best. If an NCO has appeared before an MOS Medical Retention Board and been determined fit for duty and deployable, rating officials may not state that the profile hinders duty performance. For pregnant NCOs who have not taken the APFT within the last 12 months due to pregnancy, maternity leave, and temporary profile, the rater will enter the following statement in Part Ivc: "exempt from APFT requirement in accordance with AR 40-501."

Now a list of bullet comments for each area in Parts IV and V follows so that you will have an idea as to the kind of bullet comments you may want to write. The bullet comments in this book are only examples. Using the following word lists, you can add to or take away from your own bullet comments.

Excellent and Success Adjectives and Verbs

- Accurate
- Ace
- Active
- Affirmative
- Alert
- All-around
- Appealing
- Calm
- Candid
- Capable
- Charismatic
- Clear-thinking

- Competent
- Complete
- Composed
- Concise
- Confident
- Consistent
- Contractive
- Cooperative
- Courageous
- Courteous
- Creative
- Curious
- Decisive
- Dedicated
- Dependable
- Determined
- Diligent
- Diplomatic
- Discreet
- Dynamic
- Eager
- Effective
- Efficient
- Eminent
- Enthusiastic
- Excellent
- Exceptional
- Expert
- Extraordinary
- Extreme
- Fair-minded
- Favorable
- Fearless
- Fine
- First-string
- Flexible
- Forceful
- Foremost
- Forward-looking

- Gallant
- Generous
- Genuine
- Good-humored
- Good-natured
- Gung-ho
- Helpful
- High
- Honest
- Imaginative
- Important
- Independent
- Innovative
- Intense
- Involved
- Knowledgeable
- Loyal
- Major
- Mature
- Maximum
- Meaningful
- Motivated
- Neat
- Objective
- Open-minded
- Optimistic
- Orderly
- Organized
- Original
- Outgoing
- Outstanding
- Patient
- Perceptive
- Perfect
- Persevering
- Persuasive
- Pleasant
- Polished
- Positive

- Powerful
- Practical
- Precise
- Predictable
- Productive
- Professional
- Progressive
- Proper
- Punctual
- Quick
- Rational
- Realistic
- Remarkable
- Resourceful
- Respectful
- Responsive
- Self-confident
- Self-demanding
- Significant
- Sincere
- Sizable
- Sound
- Special
- Splendid
- Stern
- String
- Successful
- Superior
- Supportive
- Systematic
- Tactful
- Thorough
- Trustworthy
- Understanding
- Unique
- Valuable
- Vigorous
- Well-liked

There are a countless number of other adjectives you could find and use, but to put them all in this book would be a book in itself. As you find more, you can write them down in this book for your use. Now, for the list of verbs.

- Accepts
- Achieves
- Accomplishes
- Accounts
- Acquaints
- Acquires
- Acts
- Actuates
- Adapts
- Adheres
- Adjusts
- Administers
- Advances
- Advises
- Advocates
- Agitates
- Analyzes
- Anticipates
- Applauds
- Applies
- Appraises
- Appropriates
- Approves
- Arises
- Arouses
- Arranges
- Articulates
- Ascends
- Aspires
- Assembles
- Asserts
- Assigns
- Assists
- Assumes

- Attains
- Attempts
- Authorizes
- Avoids
- Bolsters
- Builds
- Calculates
- Capitalizes
- Carries out
- Challenges
- Checks
- Circumvents
- Collaborates
- Commands
- Communicates
- Complies
- Comprehends
- Computes
- Conceives
- Concentrates
- Conducts
- Conforms
- Confronts
- Considers
- Consolidates
- Consults
- Contemplates
- Continues
- Contributes
- Controls
- Conveys
- Cooperates
- Coordinates
- Copes

- Creates
- Cultivates
- Dedicates
- Delegates
- Demonstrates
- Determines
- Deters
- Develops
- Devises
- Devotes
- Directs
- Discusses
- Displays
- Distinguishes
- Dominates
- Drafts
- Effects
- Embodies
- Emerges
- Emphasizes
- Employs
- Emulates
- Encourages
- Endeavors
- Energizes
- Enforces
- Enhances
- Enlightens
- Enriches
- Entices
- Erupts
- Establishes
- Evaluates
- Evidences

- Examines
- Exceeds
- Excels
- Excites
- Executes
- Exercises
- Expands
- Expects
- Expedites
- Exploits
- Explores
- Expresses
- Fabricates
- Faces
- Facilities
- Fine-tunes
- Focuses
- Follows up
- Formulates
- Fortifies
- Fulfills
- Gains
- Generates
- Gives
- Glorifies
- Grasps
- Handles
- Hastens
- Helps
- Identifies
- Ignites
- Illustrates
- Immerses
- Implements
- Imposes
- Impresses
- Improves
- Improvises
- Influences

- Informs
- Initiates
- Inspects
- Inspires
- Instigates
- Instills
- Insures
- Interacts
- Interprets
- Interviews
- Issues
- Judges
- Keeps
- Knows
- Launches
- Learns
- Maintains
- Makes
- Manages
- Manipulates
- Meets
- Motivates
- Negotiates
- Notifies
- Nourishes
- Obtains
- Operates
- Organizes
- Originates
- Outlasts
- Overcomes
- Oversees
- Overwhelms
- Paces
- Participates
- Perceives
- Performs
- Persists
- Persuades

- Plans
- Possesses
- Practices
- Prepares
- Presumes
- Prevents
- Processes
- Prods
- Produces
- Projects
- Promotes
- Proposes
- Prospers
- Provokes
- Purges
- Pursues
- Quantifies
- Quickens
- Radiates
- Rallies
- Realizes
- Receives
- Recognizes
- Recommends
- Records
- Reflects
- Regards
- Regulates
- Reinforces
- Relates
- Releases
- Relies
- Renew
- Reorganize
- Reports
- Represents
- Requires
- Respects
- Responds

- Reviews
- Revises
- Revives
- Schedules
- Secures
- Seeks
- Serves
- Shows
- Solves
- Sparks
- Spearheads
- Stimulates
- Strengthens
- Strives
- Studies
- Supervises
- Supports
- Surpasses
- Sustains
- Takes
- Thrives
- Tolerates
- Trains
- Transforms
- Treats
- Understands
- Uses
- Utilizes
- Verifies
- Weighs

There you have it. A list of words you can use to write "Excellent" and Success bullet comments. It is not hard to write a good NCO-ER. The bad ones are the hard ones to write because there are times you may want to show the NCO needs more training but, at the same time, you don't want him to be forced out of the Army. The following list of words will help you write a report for the NCO who is not doing as well as you think he/she should.

"Need Improvement" Nouns, Verbs, and Adjectives

- Adversity
- Blemish
- Chagrin
- Confusion
- Defect
- Demerit
- Demise
- Dependency
- Detriment
- Deviate
- Disadvantage
- Disappointment
- Disaster
- Discord
- Discredit
- Disorder
- Dissatisfaction
- Error
- Flaw
- Forfeit
- Friction
- Hardship
- Harm
- Hindrance
- Imbalance
- Imperfection
- Impossibility
- Impurity
- Inaction
- Inadequacy
- Inattention
- Inconsistency
- Inconvenience
- Inefficacy
- Inefficiency
- Infraction
- Lack
- Lag
- Lapse
- Liability
- Loser
- Misfortune
- Mix-up
- Nonsense
- Nuisance
- Opposition
- Oversight
- Pitfall
- Problem
- Regress
- Regression
- Restraint
- Rigor
- Shortcoming

- Excuse
- Eyesore
- Failure
- Fault
- Fizzle
- Insignificance
- Interference
- Intrusion
- Invalidity
- Irregularity
- Shortfall
- Sloppy work
- Tension
- Uncertainty

There are some very harsh words you can use to write a bad NCO-ER, so be very careful about the words you select. Remember, the NCO should always be counseled if he/she is not performing as you think he should. Now, for a list of verbs that can be used for the "Needs Improvement" rating.

- Collapses
- Conceals
- Concedes
- Condescends
- Crime
- Debases
- Denounces
- Deprives
- Destroys
- Deteriorates
- Dilutes
- Diminishes
- Disappoints
- Disrupts
- Dodges
- Dwindles
- Eludes
- Encumbers
- Erodes
- Eschews
- Exaggerates
- Extenuates
- Fades
- Fails
- Fakes
- Falters
- Flops
- Flounders
- Flunks
- Foils
- Gloom
- Hampers
- Harms
- Hinders
- Immobilizes
- Impairs
- Impedes
- Incapacitates
- Irritates
- Lacks
- Lags
- Lapses
- Lavishes
- Lessens
- Limps
- Lowers
- Mistakes
- Negates
- Neglects
- Obstructs
- Ousts
- Palls
- Quits
- Rebuts
- Refuses
- Rejects
- Relinquishes
- Restrains
- Retards
- Stagnates
- Suppresses
- Sways
- Transgresses
- Violates
- Vitiates
- Weakens
- Wears-out
- Wilts
- Works-over
- Worsens

Here's a list of adjectives that can be used to write a bad or not so good NCO-ER. Again, be careful of the words you choose.

- Abnormal
- Adverse
- Amiss
- Conflicting
- Cumbersome
- Defective
- Deficient
- Desperate
- Detrimental
- Disappointing
- Dispassionate
- Disruptive
- Effortless
- Elusive
- Erosive
- Erratic
- Erroneous
- Evasive
- False
- Faulty
- Fearful
- Feeble
- Forbidden
- Formless
- Fragile
- Fruitless
- Gloomy
- Good-for-nothing
- Grave
- Grievous
- Grim
- Gross
- Haphazard
- Hard-put
- Harsh
- Helpless
- Hit-or-miss
- Hopeless
- Humiliating

- Idle
- Ill-advised
- Ill-fated
- Ill-gotten
- Illegal
- Illicit
- Imaginative
- Imperfect
- Impossible
- Impotent
- Impractical
- Imprecise
- Improbable
- Improper
- Inaccurate
- Inactive
- Inadequate
- Inadvisable
- Inappropriate
- Incapable
- Incomparable
- Incompatible
- Incomplete
- Incomprehensible
- Inconclusive
- Inconsiderable
- Inconsistent
- Inconvenient
- Incorrect
- Indefensible
- Ineffective
- Inefficient
- Ineligible
- Inert
- Inexact
- Inexcusable
- Inferior
- Inopportune
- Inordinate

- Insecure
- Insignificant
- Insoluble
- Insubstantial
- Insufferable
- Insufficient
- Insupportable
- Intolerable
- Intolerant
- Intractable
- Intrusive
- Invalid
- Irregular
- Irrelative
- Irrelevant
- Last
- Last-ditch
- Limited
- Lost
- Low-grade
- Low-level
- Ludicrous
- Meaningless
- Miserable
- Negative
- Negligent
- Nonproductive
- Null
- Obscure
- Obsolete
- Out-of-date
- Outcast
- Outlandish
- Outrageous
- Overdue
- Pathetic
- Petty
- Problem
- Purposeful

- Redundant
- Resistless
- Rigorous
- Rough
- Run-down
- Rusty
- Scant
- Shabby
- Skeptical
- Skepticism
- Sloppy
- Small-scale
- Somber
- Sorrowful
- Sparse
- Spotty
- Stagnant
- Subnormal
- Substandard
- Thriftless
- Tiresome
- Tricky
- Uncertain
- Uneasy
- Unfavorable
- Valueless
- Wanting
- Weariful
- Wearisome
- Washed-up
- Wasteful
- Weak

Now you have a small list of words that you can use to write your bullet comments. Make sure your comments justify the rating. If you should receive an NCO-ER and feel that the bullet comments do not justify the rating, let your rater or senior rater know. At the same time, you can write down two or three that you may want him/her to use. The best way to do this is to write down two or three bullet comments that you would like to get on your report, then write down two more that you know are more than should be said about your rating. Most likely, the rater will pick the same ones you wanted.

Now, I am going to give you a list of bullet comments that might help you relate to the one you are writing. You will find comments that you can use in all rated areas, which are:

- Values
- Competence
- Physical Fitness and Military Bearing
- Leadership
- Training
- Responsibility and Accountability
- Overall Performance
- Overall Potential for Promotion

Under each heading, I will give you a list of bullet comments that can be used for the "Excellent," "Success" or "Needs Improvement" rating. You should be able to tell which comment is for what rating. Remember, the "Excellent" rating is given to few NCOs. The "Success" rating is something that most NCOs receive, and the "Needs Improvement" rating is for the NCO who is not doing something right. Let's get started by examining different types of bullet comments.

Bullet Comments for Values

Positive Values Bullet Comments

- supports military and civilian chain of command
- places unit needs and goals first
- does what is needed without being told
- accepts full responsibility of soldiers and self
- is willing to sacrifice
- honest, upright, sincere, and candid
- avoids deceptive behavior
- does the right and moral thing
- is a good role model
- develops self and soldiers
- soldiers ethically
- shares thought process with soldiers
- chooses action that best serves the nation
- unparalleled loyalty
- is loyal to organization
- displays absolute loyalty to superiors
- builds loyalty in soldiers
- places unit's interest ahead of own
- is committed to unit goals
- displayed loyalty during "Operation Desert Storm"
- extremely dedicated to the cause
- takes pride in job performance
- displays loyalty in profession
- effectively handles pressure
- effectively manages stress
- makes positive use of stress
- performs well under pressure
- performs well under stress

- performed well during "Operation Desert Storm"
- works effectively in high-pressure situations
- gets the job done
- gains control over job pressures
- keeps growing professionally
- great personal drive
- keen sense of ethical conduct
- admirable courage
- never gives up on mission
- honest and faithful
- stands behind principles
- proper personal behavior
- high ethical principles
- great personal behavior
- shows courage under pressure
- composed under pressure
- engaging personality
- never loses temper
- sound character
- friendly and sociable
- personal devotion to duty
- devoted to Army ethics
- genuine concern for others
- high concern for others
- caring concern for others
- impeccable, winning spirit
- promotes loyalty
- unselfish and trusting
- strength of character
- has high ideals, morals, and ethics
- unshakable character
- faces fear head-on
- ethical, honest personality
- strong ethical principles
- great sense of loyalty

- continually exhibits loyalty
- keeps composure under fear and stress
- strong advocate of Army ethics
- not easily excited under pressure
- not easily excited under stress
- possesses highest personal loyalty
- unbearable character and loyalty
- high morality and ethical principles
- even, steady temperament
- calm and composed during combat
- has high personal values
- unafraid of combat challenge
- positive unit spirit
- strong loyalty and sense of pride
- completely without bias or prejudice
- actively promotes soldiers' rights
- encourages professional pride
- encourages subordinates' loyalty
- instills loyalty and pride
- contributes 110 percent to team effort
- promotes harmony and team work
- leader and member of the team
- encourages excellence among soldiers
- fosters high trust level in soldiers
- dedicated to unit mission
- mission first, self last
- treats others with respect
- willing to share knowledge with soldiers
- demands maximum effort in team efforts
- stable and calm during "Operation Desert Storm"
- establishes team spirit throughout
- highly dedicated to unit goals
- devotion to duty and self-sacrifice
- a true team leader
- displays selfless devotion to duty

- unsurpassed devotion to duty
- unsurpassed devotion to soldiers' care
- self-sacrificing, a true team player
- has longstanding record of loyalty
- will not retreat in the face of fear

Negative Values Bullet Comments

- causes disorder and unrest
- disagreeable personality
- misrepresents the facts
- uncaring in matters of sympathy
- disruptive influence
- verbally abuses soldiers
- wears down subordinates' spirit
- disobedient and belligerent
- incompatible, disagreeing personality
- incompetent without direct, consistent supervision
- unmanageable off-duty activities
- inclined to stray from the truth
- mood change without warning
- low level of self-confidence
- lacking in social graces and courtesy
- gives into pressure during crisis situations
- cannot make proper decisions under pressure
- prejudicial to good order and discipline
- substandard duty performance
- unwillingness to conform to Army ethics
- overly apologetic and humble
- opposes all views other than own
- arbitrarily enforces standards
- lack of self-confidence
- requires close supervision
- fabricates the truth
- uncaring team leader
- cannot deal with stress and fear

- unwilling to lead during "Operation Desert Storm"
- shies away from soldiers' problems
- no pride in self or duty performance
- places self ahead of duty, unit and Army
- uncommitted to unit goals
- lackadaisical attitude
- short-tempered with soldiers
- untruthful to soldiers and leaders
- uninterested in Army ethics
- becomes insubordinate when talking to leaders

Bullet Comments for Competence

Positive Competence Bullet Comments

- demonstrates competent performance
- projects a special competence
- displayed competency during "Operation Desert Storm"
- demonstrates a high level of expertise
- demonstrates strong personal effectiveness
- demonstrates strong interpersonal competence
- excels in the effective application of skills
- displays a high level of technical competence
- blends technical skills with technical competence
- combines technical competence with loyalty
- demonstrates highly sophisticated skills
- processes specialized teaching skills
- processes specialized writing skills
- processes specialized speaking skills
- processes specialized listening skills
- processes specialized learning skills
- highly skilled in all phases of job
- highly skilled in all phases of training
- highly skilled in all phases of counseling
- displays excellent attention to technical skills

- is uniquely qualified to teach
- effectively capitalizes on strengths
- constantly sharpens and updates skills
- demonstrates professional expertise
- can be relied on to make sound decisions
- is willing to make unpopular decisions
- assembles facts before taking actions
- weighs alternative decisions before taking actions
- carefully evaluates alternative risks
- practices sound risk taking
- foresees the consequences of decisions
- makes sound decisions under pressure
- avoids hasty decisions
- excels in profit-directed decisions
- concentrates on developing solutions
- excels in seeking solutions
- makes decisions with confidence
- strives to improve decisiveness
- encourages decision with confidence
- demonstrates consistently distinguished performance
- ability to achieve desired results
- achieves bottom-line results
- attains results through positive actions
- produces a tangible, positive impact
- exceeds performance expectations
- performance exceeds job requirements
- provides a competitive edge
- turns risk situations into opportunities
- possesses all traits associated with excellence
- uses all available resources
- generates enthusiasm
- is extremely resourceful
- works diligently until job completion
- displays concentrated effort
- displays trust and confidence

- faces conflicts with confidence
- completes tasks with confidence
- extremely self-confident
- maintains a high degree of involvement
- develops positive expectations
- develops realistic training
- sets high standards of personal performances
- writes in a positive tone
- writes reports that achieve maximum impact
- excels in making appropriate judgments
- can be entrusted to use good judgment
- effectively solves problems
- translates problems into workable solutions
- ability to solve problems
- diagnoses situations or conditions
- considers alternative courses of action
- exercises judgment on behavior of others
- shows eagerness and capacity to learn
- ability to learn rapidly
- benefits from all learning situations
- consistently strives to improve performance
- excels in self-supervision
- excels in self-improvement
- thirst for knowledge
- sound judgment
- quick to learn
- interesting, convincing speaker
- advanced knowledge
- extensive knowledge
- firm, caring attitude
- exercises sound judgment
- capable of independent decision
- unlimited learning capacity
- relentless drive and ambition
- persuasive talker

- matchless desire
- self-motivated
- results-oriented
- articulate speaker
- thinks and plans ahead
- eager and capable
- creative writing ability
- commands large vocabulary
- sound in thought, good in judgment
- skilled, eloquent speaker and writer
- unmatched appetite for learning
- strong ability to learn
- polished, persuasive speaker
- inspires self-improvement in soldiers
- always produces quality results
- learning growth potential
- unending drive to win
- unending drive and urge for success
- unquenchable thirst for knowledge
- continually seeks personal growth and development
- capable of independent thought and action
- has "follow me" confidence
- made sound judgments during "combat training"
- winning attitude during "Operation Desert Storm"
- set standards during "Operation Desert Storm"

Negative Competence Bullet Comments

- poor planner, great hindsight
- lacking in knowledge
- lax in performance and behavior
- lacks in desire
- low self-esteem
- bad judgment
- serious judgment error

- shuns duty
- unpredictable work habits
- mild learning disability
- persistently poor performance
- does no more than required
- lack of pride in work
- lack of initiative
- negative attitude
- lack of pride in work
- exercises bad judgment
- fails to monitor subordinates
- lacks self-discipline
- sloppy workmanship
- frequent bad judgment
- gives misguided direction
- obvious lack of self-motivation
- gets less than required results
- fails to achieve consistency
- lack of confidence in abilities
- lacks knowledge and ability
- lacks consistency in performance
- deficient in skill and knowledge
- incapable of sustained satisfactory performance
- uncertain in making decisions
- doubtful in making decisions
- lacks good judgment and common sense
- unwilling to listen to soldiers
- avoids duty during prime time training
- written products are not clear and coherent
- blunt and rude in speech and manner
- unable to choose correct course of action
- shows no desire for improvement
- exhibits only short periods of success
- unable to distinguish right from wrong
- decisions are open to question

- fails to achieve minimum acceptable performance
- makes careless, avoidable mistakes
- opposes improvement efforts of others
- cannot articulate in speech
- incoherent in writing
- dodges work and responsibility
- not job-aggressive
- relies too heavily on others
- below-average performer
- sometimes premature in judgment
- best efforts prove fruitless
- best efforts prove unsuccessful
- speaks and acts on impulse
- plagued by lack of self-confidence
- inattentive to details
- slow and deliberate work pace
- performance is all valleys and no peaks
- performs well below peer group
- less than marginal performer
- seriously lacking in initiative
- lacks counseling skills
- total lack of enthusiasm
- reluctance to conform to standards
- unsophisticated reasoning and judgment
- slow to learn and develop
- failed to achieve consistency in training
- lacks the capacity to lead
- bad leader, good follower
- loses composure when under stress
- unsuitable for NCO corp
- sometimes lax in supervision
- not serious about training soldiers
- has no idea how to train soldiers
- unable to train soldiers
- CTT scores lower than his soldiers

- has difficulty with training soldiers
- planned badly for SDT
- makes hasty decisions

Bullet Comments for Physical Fitness and Military Bearing

Positive Physical Fitness and Military Bearing Comments

- grasps the most difficult concepts
- exceptionally keen and alert
- reasonable, smart, and keen
- alert, quick, and responsive
- sustains a high level of concentration
- independent thinker
- thinks before taking action
- thinks fast on feet
- uses common sense
- uses sound fact-finding approaches
- demonstrates innovative insight
- uses intelligent reasoning
- displays strong memory skills
- displays strong mental flexibility
- possesses strong memory skills
- thinks futuristically
- effectively manages stress
- performs well under pressure
- gets things done calmly
- recognizes the importance of appearance
- presents an attractive appearance
- takes pride in personal appearance
- grooming is neat, attractive, and appropriate
- master at personal hygiene and dress
- conforms to proper standards of dress

- dresses appropriately for all occasions
- projects a positive image
- perceptive and alert
- composed and calm
- great mental grasp
- a quick thinker
- always enthusiastic
- emotionally stable
- mentally alert
- mentally sharp
- bold, forward thinker
- calm and affable manner
- powerful, influential figure
- alert, energetic personality
- sound of mind and judgment
- pleasing personality
- analytical mind
- boundless energy
- intellectual courage
- strong will of mind
- sharp, mental approach
- quick, penetrating mind
- a pillar of strength
- 10K running team member
- post basketball team member
- mentally and physically able
- scored 300 on APFT
- another 300 APFT score
- only E-7 on post to score 300 on APFT
- awarded the Army Physical Fitness Patch
- outscored all peers in BN on APFT
- outscored his/her platoon during APFT
- coach for unit football team
- in charge of overweight soldiers in unit
- teaches stress management to peers
- unit stress management NCO

- organized unit "quit smoking" campaign
- organized training for overweight soldiers
- uses common sense to tackle problems
- exceptional personal drive and energy
- appearance, without equal
- positive mental attitude and outlook
- mentally quick and active
- strong, positive drive
- a pillar of moral strength and courage
- dignified in presence and appearance
- physically ready, mentally alert
- mannerly, courteous, and polite
- displayed physical vigor during BN "Organization Day"
- strong will of mind and commanding presence
- impressive posture and appearance
- meets Army standards by APFT
- always active, on or off duty
- great mental aptitude
- a perfect example of physical fitness

Negative Physical Fitness and Military Bearing Comments

- not physically fit by Army APFT standards
- lacks physical vigor
- weak, inadequate leader
- not well-organized mentally
- emotionally immature
- failed last APFT
- careless appearance
- lacks military bearing
- failed to improve low APFT scores
- lack of desire for physical training
- loses emotional control
- lacks proper mental discipline
- does no more than required during APFT

- sets bad example for physical fitness
- lags behind others during PT run
- failed to live up to expectations of a leader
- lacks mental depth and soundness
- not worthy of present rank
- sometimes lax in physical training
- sometimes lax in appearance
- military bearing is below standards
- substandard military bearing
- has bad habit of skipping morning PT

Bullet Comments for Leadership

Positive Leadership Bullet Comments

- performs with a high degree of accuracy
- performs with consistent accuracy
- strives for perfection
- excels in achieving perfection
- believes in bottom-up leadership
- meets precise standards
- communicates high expectations
- communicates clearly and concisely
- demonstrates sound negotiating skills
- creates opportunities
- initiates fresh ideas
- is very dependable and conscientious
- extremely reliable and supportive
- successfully builds soldiers
- trains soldiers to become leaders
- facilitates learning
- builds on strengths
- identifies soldiers' needs
- plans for effective career development
- seeks personal growth and development
- understands personal strengths and weaknesses

- encourages broad development of soldiers
- assists soldiers with skills, knowledge, and attitudes
- sets reachable targets
- sets reachable goals
- sets worthy goals
- is a goal seeker
- effectively establishes group goals
- achieves and surpasses goals
- shows concern for soldiers
- believes in soldiers' care
- shows concern for soldiers' development
- concern about professional development
- created study plan for NCO's SDT
- takes charge without being told
- always ready to take over the next leadership position
- radiates confidence
- inspires confidence and respect
- projects self-confidence
- demonstrates natural leadership ability
- demonstrates strong, dynamic leadership
- ability to stimulate others
- commands the respect of others
- knows how to get soldiers' attention
- maintains a mature attitude
- speaks with a positive tone
- speaks with a pleasant tempo
- excellent persuasive abilities
- develops sound contingency plans
- develops positive expectations
- expects and demands the best
- committed to excellence
- strives for perfection
- performs at peak performance
- performs at a high energy level
- exceeds normal output standards
- understands human behavior

- easily gains acceptance of others
- best impression in every situation
- conveys a positive personal image
- gets along well with others
- builds a close rapport
- promotes harmony among soldiers
- interacts effectively with soldiers and peers
- conveys positive influences
- excels in promoting team efforts
- establishes realistic work demands
- encourages soldiers to win
- effectively balances workflow
- builds cooperation
- promotes positive involvement
- stimulates individual participation
- gives helpful guidance to new soldiers
- shows a sincere interest in soldiers
- excels in effective counseling of soldiers
- effectively uses counseling techniques and skills
- gives sound, practical advice
- properly asserts authority
- gains soldiers' confidence
- shows empathy
- shows genuine respect
- sensitive to the feelings of others
- supervises firmly and fairly
- is fair and firm with soldiers
- turns complaints into opportunities
- quickly settles disciplinary problems
- corrects without criticizing
- takes appropriate remedial action
- tactful in conflict situations
- negotiates with tact
- fighting enthusiasm
- always ahead of the action

- good samaritan
- dedicated to helping others
- dedicated to soldiers' care
- result-oriented individual
- unlimited capacity for solving problems
- highly motivated achiever
- always a willing volunteer
- motivates and leads others
- a real motivator
- composed leader
- good organizer
- firm, resolute leader
- bolsters spirits
- inspires performance
- good-natured team leader
- positive motivator
- firm, yet fair leader
- respected by peers and other leaders
- uncommon leadership
- inspires greatness
- promotes sound leadership
- accomplished counselor
- skillful, direct leadership
- astute, experienced leader
- unparalleled leadership
- motivates and leads others
- well-rounded leadership skills
- stirs up enthusiasm
- charismatic leader
- no-nonsense leader
- inquisitive leader
- impressive leadership
- selfless leader
- team leader
- concerned, caring leader

- has personal leadership magic
- sensitive to the needs of others
- frank, direct leader
- an accomplished counselor
- vigorous leadership style
- promotes *esprit de corps*
- impressive leadership record
- tactful leader and motivator
- engenders self-development
- exercises sound leadership principles
- exceptional leadership
- influence soldiers to accomplish the mission
- provides purpose, direction, and motivation
- correctly assesses soldiers' competence
- correctly assesses soldiers' motivation
- correctly assesses soldiers' commitment
- takes proper leadership action
- creates climate that encourages soldier participation
- develops mutual trust, respect, confidence in soldiers
- is aware of his/her strengths, weaknesses, and limitations
- ensures soldiers are treated with dignity and respect
- considers the available resources
- considers subordinates' level of competence
- skillful in identifying/thinking through the situation
- takes quick corrective action
- seeks self-improvement
- technically and tactically proficient
- seeks and takes responsibility
- makes sound and timely decisions
- leader in soldiers' care
- keeps soldiers informed
- develops a sense of responsibility in soldiers
- ensures tasks are understood, supervised, accomplished
- develops highly effective team
- knows limitations and capabilities of soldiers

- accomplishes all assigned tasks to the fullest
- willingness to accept full responsibility
- honest and upright
- sincere, honest, candid, and avoids deceptive behavior
- explains the why of the mission
- teaches soldiers to think creatively
- solves problems while under stress
- able to plan, maintain standards, and set goals
- supervises, evaluates, teaches, coaches, and counsels
- rewards soldiers for exceeding standards
- serves as ethical standards bearer
- develops cohesive soldier teams
- corrects performance not meeting standards
- leader with strong and honorable
- a role model for all to follow
- committed to Army ethics
- chooses the right course of action
- frank, open, honest, and sincere
- competent and confident leader
- establishes broad categories of skills and knowledge
- able to understand and think through a problem
- ability to say the correct thing at the right moment
- does not interrupt when others are speaking
- ensures soldiers are professionally and personally developed
- ensures the tasks are accomplished
- influences the competence and confidence of soldiers
- helps soldiers develop professionally and personally
- understands how soldiers learn
- motivates soldiers to learn
- an expert in teaching and counseling
- counsels soldiers frequently
- counsels soldiers on strengths and weaknesses
- good at identifying and correcting problems
- understands human nature
- has great listening skills
- leads soldiers to making their own decisions

- knows when to be flexible and unyielding
- creates strong bond with soldiers
- builds soldiers' spirit, endurance, skills, and confidence
- works hard at making soldiers team members
- knows war-fighting doctrine
- operates and maintains all assigned equipment
- not afraid to ask seniors, peers, or soldiers for help
- understands doctrine and tactics of potential enemies
- gathers facts and makes assumptions
- develops possible solutions
- analyzes and compares possible solutions
- recognizes and defines the problem
- selects the best solutions
- able to make timely decisions
- identifies the best course of action
- involves soldiers in making decisions
- understands and uses backward planning
- develops a schedule to accomplish task
- recognizes and appreciates soldiers' abilities
- uses computers, analytical techniques, and other technological means
- good example setter
- corrects performance not meeting standards
- makes soldiering meaningful
- makes decisions soldiers accept
- plans and communicates effectively
- an example of individual values
- learns leadership skills from other leaders
- teaches soldiers to depend on each other
- develops full potential of soldiers
- constantly stresses teaching, coaching, and caring
- constantly stresses bonding, learning, and teaching
- creates trust and strong bonds with and among soldiers
- strong desire to excel
- inspires zeal and obedience
- displays concern for soldiers

- experienced, knowledgeable leader
- ability to inspire others
- stimulating leader
- a "leader by example" who obtains superior results
- a dynamic, motivating leader who obtains superior results
- strict, firm disciplinarian
- animating leader, arouses enthusiasm
- supervises and directs soldiers
- instills pride and dignity in others
- strong, decisive leader
- dynamic leader and vigorous worker
- dedicated to unit's mission and goals
- hard-line leader with unyielding character
- provides assistance to those in need
- potent, productive leader
- vigorous leadership, superb management
- capitalized on soldiers' strengths
- epitome of tactful leadership
- stimulates subordinates' professional growth
- intense, compassionate leader
- invigorating, successful leader
- brings out best in soldiers
- gives support to chain of command
- informed leader who cares about soldiers
- compassionate, caring leader
- considerate of the feelings of others
- shares time and knowledge with soldiers
- offers positive advice to soldiers
- a demanding leader who gets impressive results
- demonstrates superb leadership
- work-aggressive leadership
- radiant, confident leader
- highly respected leader and organizer
- superb leadership
- master of superb leadership
- proven leader of unbounded ability

- imaginative leadership techniques
- poised and mature leader
- assertive and considerate leader
- proponent of strong, solid leadership
- persuasive and tactful leader
- leads with intensity, force, and energy
- develops soldiers at a rapid pace
- encourages off-duty professional growth
- enforces rules and regulations
- personifies leadership by example
- gives a pat on the back when earned
- knows key to quality performance
- recognizes and rewards top performances
- ignites enthusiasm throughout the ranks
- effective leadership qualities
- unfailing devotion to duty and country
- gives praise when most deserved or needed
- equal and equitable treatment of soldiers
- a real motivator and dedicated leader
- encourages each soldier to set high goals
- leads soldiers to desired level of performance
- demands the best from soldiers
- displays tactful leadership daily
- combat leader
- gives subordinates demanding responsibility
- leader of uncommon perceptiveness
- stands above peers
- high achiever
- unblemished record
- walking FM 22-100
- a role model for all NCOs accustomed to success
- a top professional
- committed to excellence
- seeks challenging assignments
- sets professional example

- overcomes all obstacles
- superior to peers
- head and shoulders above peers
- reliable and dependable
- skilled in art of leadership
- highly specialized in leadership
- made marked improvement during past year
- a champion in the field of leadership

Negative Leadership Bullet Comments

- lacking in knowledge
- dispassionate leader
- weak, inadequate leader
- impersonal leader
- inferior performer
- inconsiderate and uncaring
- dull, uninspiring leader
- not a potent, effective leader
- weak, ineffective leader
- lack of confidence
- leadership vacuum
- insensitive leadership
- unimpressive leader
- inflexible leader
- concerned with own self-interest
- lags behind others
- apathetic leader
- suppresses subordinates' growth
- inflexible, unimaginative leadership
- deficient in skills and knowledge
- failed to live up to expectations
- irresponsible leadership
- impatient with subordinates
- bad leader and worse follower

- insensitive and calloused leader
- not a strong leader
- too permissive and lenient
- careless of the feelings of others
- suitable for routine, ordinary jobs
- unforgiving leader, rubs in mistakes by subordinates
- leadership style leaves uneasy feeling
- non-forgiving leader, doesn't forget others' mistakes
- leadership ability is questionable
- total lack of enthusiasm
- loses composure when under pressure
- stirs up resentment
- over-supervises soldiers
- breeds resentment and low morale
- lack of coordination
- disorganized leader
- sets bad examples for soldiers
- a leader that should not be
- can't lead more than one soldier
- wants credit but can't lead
- a leader with no know-how
- last NCO to leave in charge
- can't be trusted to accomplish the mission
- soldiers tell him/her what to do

Bullet Comments for Training

Positive Training Bullet Comments

- a great raw ability and talent
- stimulating intelligence
- powerful, influential figure
- advanced knowledge
- unlimited teaching capacity
- sparks excitement
- personal devotion to training

- winning spirit
- trains to changing situations
- dedicated to helping soldiers learn
- teaches from combat experience
- realistic trainer
- superb trainer and leader
- unmatched capacity for teaching
- strong desire and ability to teach and train
- persuasive trainer and supervisor
- trains soldiers to be trainers
- teaching and training ability is unlimited
- always coming up with something better
- stands above contemporaries in training soldiers
- always willing to teach and train
- most professional talent is training
- a leader in training soldiers
- a take-charge attitude with great training ability
- shares knowledge with others
- possesses an impressive breadth of experience
- continually strives for professional development
- promotes professional growth
- plans ahead for training success
- persuasive, convincing trainer
- strong will and know-how
- relentless drive and dedication
- excellent teaching ability
- uses own initiative to train soldiers
- possesses abundance of enthusiasm and drive
- stands up for principles and beliefs
- devotion to duty always
- teaches soldiers how to survive during combat
- real morale booster and trainer
- trains soldiers to win and survive
- arouses and excites when training
- accomplished trainer
- skillful, direct trainer and leader

- inspires and encourages during training
- well-rounded training skills
- stirs up enthusiasm
- no-nonsense trainer
- impressive trainer
- trains so all can learn
- fully taxes soldiers when training
- inspires the imagination
- positive influence, great motivator
- knows success in team training
- vigorous training style
- impressive training record
- enforces soldiers' participation
- has ability to inspire others to train
- experienced, knowledgeable trainer
- dedicated to betterment of subordinates
- contributes maximum effort and energy
- invigorating trainer and leader
- excites and arouses others to action
- training style elicits harmony and team work
- dedicated to mission training
- brings out the best in subordinates
- effectively trains and directs inexperienced personnel
- skillful in training others to desired goals
- training merits special praise and gratitude
- performance-oriented trainer
- trains to sustain proficiency
- trains to maintain
- trains to achieve combat-level standards
- trains soldiers to fight
- effectively uses realistic conditions when training
- includes simulations, simulators, and devices in training
- trains to challenge
- his/her training excites and motivates soldiers
- sustains skills to high standards
- operates in a "band of excellence"

- uses all available time for training
- responsible for training and performance of soldiers
- bases training on wartime mission requirements
- provides the required resources for training
- helps develop mission training programs
- awarded for best mobilization plan
- responsible for the best maneuver team
- develops supporting task list for each METL
- maintains consistent training awareness
- concerned with future proficiency
- effectively uses all available resources
- allocates maximum training time
- provides training recommendations for unit training
- helps improve soldiers' war fighting skills
- provides specific guidance to trainers
- prepares detailed training schedules
- creates a bottom-up flow of information regarding training
- specifies tasks to be trained
- provides concurrent training topics for training
- specifies who conducts and evaluates training
- provides training feedback to commander
- personally observes and evaluates all training
- demands training feedback from subordinate leaders
- executes and evaluates training
- coaches trainers on how to train
- provides time to rehearse training
- competes to achieve the prescribed standards
- trains evaluators as facilitators
- creates well-trained and highly motivated soldiers
- is honest about wartime situations
- guides team towards mission accomplishment
- listens and responds fairly to criticisms
- guides soldiers into accepting team goals
- sets realistic goals
- periodically checks on team progress and training
- concerned about soldiers' safety and survival in combat

- a catalyst of team work and high morale
- willing to share knowledge and talent
- promotes maximum effort in team training
- first to learn, last to forget
- a proven trainer of unbounded ability
- bold and imaginative training techniques
- superior knowledge of training and teaching
- always willing to train others
- persuasive and tactful trainer
- trains with intensity, force, and energy
- realistic, motivated trainer
- spent extensive hours developing realistic training plans
- never too busy to train soldiers
- unique ability to coordinate group efforts
- popular among peers and soldiers
- molds team into cohesive, productive unit
- establishes and enforces clear-cut goals
- demands positive results
- successful conclusion never in doubt
- trains team members to take charge
- trains team members to train each other
- true team player
- plans carefully and wisely
- trainer of trainers
- primary trainer for unit
- trains to win and succeed
- trains and tests results

Negative Training Bullet Comments

- dull, uninspiring trainer
- unable to master training skills
- loses control when training
- lack of confidence when training
- ignores reality
- pressures soldiers when training

- throws weight around when training
- wastes training opportunities
- deviates from trainer standards
- unimpressive trainer
- dull, trite personality
- undermines morale
- irritates and annoys others
- discourages team unity
- his/her soldiers are better trainers
- becomes easily frustrated
- lack of initiative
- fails to monitor training
- lack of personal conviction
- blames others for own shortcomings
- impatient with soldiers
- fails to achieve minimum performance
- fails to train soldiers during prime time
- lacking in ability to train
- bad habits include not training soldiers
- shies away from training
- depends on others to train his/her soldiers
- takes shortcuts in training
- his/her training leads to failure

Bullet Comments for Responsibility and Accountability

Positive Responsibility and Accountability Bullet Comments

- knows whereabouts of soldiers at all times
- accepts responsibility of own decisions
- accepts responsibility of soldiers' actions
- accepts ultimate responsibility
- assumes responsibility for mistakes
- never runs from responsibility

- effective in assigning responsibility
- delegates responsibility effectively
- devotes attention to responsibility
- meets responsibilities
- fulfills all commitments
- meets all schedules and deadlines
- can be counted on to achieve positive results
- consistently punctual
- effectively follows up assignments
- responsible leadership
- takes responsibility for right or wrong
- takes responsibility for good or bad
- understands and takes responsibility
- exceptional, responsible leader
- responsible for team morale, strength, and courage
- responsible and enthusiastic leader
- takes responsibility for soldiers' safety
- enforces responsibility and accountability
- seeks responsibility for self and soldiers
- responsible for team success
- prompt and responsive
- responsibilities discharged superbly without fail
- actively seeks additional responsibility
- thrives on important responsibility
- appetite for increased responsibility
- carries out responsibilities in a professional manner
- accepts added responsibility without wavering
- aggressive in assumption of additional responsibility
- gives wholehearted support to accountability
- assumed expanded responsibilities
- directly responsible for four-million dollars worth of equipment
- responsible for upgrading soldiers' CTT scores

Negative Responsibility and Accountability Bullet Comments

- not aware of soldiers' whereabouts
- lost valuable equipment during FTX
- shirks responsibility
- avoids work and responsibility
- didn't know equipment was missing
- will not stand up for soldiers
- does not seek responsibility
- can't count on to lead or train
- not responsible
- not accountable
- must do one thing at a time
- can't count on to care for soldiers
- wasn't aware of missing equipment
- wasn't aware his/her soldier was AWOL
- late for duty more than his/her soldiers
- forgot soldier was waiting for his/her inspection

Bullet Comments for Overall Potential for Promotion

Positive Overall Potential for Promotional Bullet Comments

- promote now
- should have been promoted yesterday
- don't hold back promotion
- promote before peers
- on the road to CSM
- sergeant with first sergeant potential
- promotion will benefit the Army
- has CSM potential
- promote yesterday
- promote ahead of peers
- highest recommendation for promotion

- whatever it takes, promote
- promotion will justify performance
- uniquely qualified for CSM
- on the road to the top
- been looked over too long, promote
- none better for next higher grade
- consistently strives to improve performance
- excels in self-improvement
- should serve as first sergeant
- promotion overdue
- look over TIS and promote
- plans for effective career development
- makes valuation suggestions for improvements
- sets ambitious growth goals
- displays an eagerness to improve
- welcomes opportunities for improvement
- high degree of professional excellence
- a self-starter
- finds new and better ways to perform duties
- does things without being told
- requires no supervision
- excels in solving critical problems
- well-informed on job and social issues
- always ready to take charge
- extremely dedicated and success-oriented
- displays intense involvement
- work in a higher grade position
- works in an E-__ position
- present position calls for an E-__
- gives maximum effort
- strong achievement drive
- maintains self-motivation
- consistent production of high quality of work
- highest standards of excellence
- quality of work reflects high professional standards
- performs at peak efficiency

- reads at a CSM level
- writes at a CSM level
- completed speed-reading course
- promotes team efforts
- seeks the hard assignments
- expects and demands superior performance
- gains maximum productivity from soldiers
- develops a spirit of teamwork
- outperforms peers three to one
- seeks hardship assignments
- optimistic outlook and attitude
- good academic aptitude
- reaches success by self-motivation
- fully capable of the next higher grade
- leadership ability is unlimited
- special talent for areas of next grade
- supervises firmly and fairly
- takes charge, a true leader and motivator
- dynamic leader with combat experience
- contributes 110 percent
- leadership merits special praise and gratitude
- leadership merits promotion to the next higher grade
- reward superior performance with promotion
- not promoting him is giving away stripes
- knows key to quality performance is leadership
- acts decisively under pressure
- provides timely advice and guidance
- sets standards by which excellence is measured
- dedicated to mission purpose
- exceptionally well-organized
- succeeds despite adversity
- self-sacrificing, duty first
- a model for all to emulate
- seeks opportunities to grow professionally
- gives serious and determined effort
- almost infinite growth potential

- well-rounded and professionally knowledgeable
- possesses all attributes required to excel to the top
- supports and enforces all rules and orders
- promptly executes all orders
- sets and achieves high goals
- realizes success requires sacrifice and dedication
- always involved in something constructive
- especially strong in the execution of demanding tasks
- did a masterful job during_____
- a foremost authority on combat training
- without equal in ability to lead
- a top specialist in the field of_____
- stands out in the field of_____
- surpasses peers in sheer ability to_____
- has the knowledge and competence to_____
- widely respected for ability to_____
- rejuvenated and put new life into training
- exhibits all essential features of a CSM
- masterful ability to lead junior NCOs
- a remarkable, skilled NCO
- impressive accomplishments include training
- rare, extraordinary ability to train junior leaders
- skilled in art of counseling
- thriving force behind success
- thoroughly understands soldiering
- maintains sharp edge in leading others
- an absolute master at training leaders
- uniquely skilled to get things done
- has veracious appetite to be the best
- no end to potential for promotion

Negative Promotion Statements

- professionally stagnated
- do not promote
- candidate for QMP

- promote after peers
- make last on promotion list
- needs to earn present rank
- promotion would be a gift
- not ready for next stripe
- working towards promotion
- hold promotion until next year
- should be reduced, not promoted
- put promotion on hold
- totally unconcerned about more responsibility
- lack of desire
- lags behind others in his/her MOS
- should move back, not forward
- no future potential
- lax in carrying out duties
- a burden to leadership
- unable to overcome minor problems
- devoid of hope for improvement
- expects success using guesswork and conjecture
- may come around in due time
- does not have the ability to lead
- not qualified to be an E-___, let alone an E-___
- incompetent in area of _____ planned badly for career progression
- has not kept pace with peers
- please don't promote
- promotion would be a waste
- if no way out, promote
- save promotion for new peers

Bullet Comments for Overall Performance

Positive Overall Performance Bullet Comments

- achieves bottom-line results
- attains results through positive actions

- produces tangible, positive actions
- performance exceeds job requirement
- provides a competitive edge
- possesses all traits associated with excellence
- displays many areas of strength
- is extremely resourceful
- works diligently
- displays strong perseverance
- competes with confidence
- high degree of involvement
- high standards of personal performance
- possesses personal magnetism
- quality of work is consistently high
- excels in getting work done by others
- expects and demands superior performance
- effectively uses counseling techniques and skills
- effectively deals with misunderstandings
- accomplishes results without creating friction
- concentrates on area in need of help
- demonstrates competence in many areas
- ability to perform a wide range of assignments
- effectively handles special assignments
- copes with accelerating changes
- totally committed to excellence
- intense dedication to duty
- stands above contemporaries
- stands above most seniors
- positive, fruitful future
- highest standards of excellence
- seeks challenging hardship assignments
- overcomes all obstacles
- ace technician and leader
- superior to others
- impressive accomplishments
- quick to take positive actions
- reaches full potential quickly

- total, complete professional
- uncommon excellence
- expects total effort
- highly competent and dedicated
- demands quality performance from all
- true meaning of "pride and professionalism"
- thrives on responsibility
- exceeds highest professional standards
- a front-runner in every category
- considers no job too difficult
- trains for the unexpected
- a leader of his peers
- performs beyond professional abilities
- at the pinnacle of professional excellence
- a zealot in performing any assignment
- takes pride in doing best job possible
- symbolizes the top quality professional
- sets standards by which excellence is measured
- executes tasks expediently and correctly
- enhances and improves morale
- impressive record of accomplishments
- sets and achieves high personal standards
- performance exceeds job requirements
- highly skilled in all phases of job
- always productively employed
- unfailing performance to duty
- achieves quality results
- puts forth unrelenting effort
- workload is correctly balanced and prioritized
- exceptionally high level of performance
- sets the pattern and example for peers
- deeply devoted to chosen profession
- frequently seeks out expert opinion
- first-rate professional
- rates first against any competition
- at the forefront of peer group

Negative Overall Performance
Bullet Comments

- downward turn in performance and attitude
- substandard performance across the board
- overall performance is below standard
- requires close and constant supervision
- does not promote good morale
- average leadership skills
- failed SDT
- performance declined during rating period
- goes out of way to irritate others
- a complete disappointment
- ignores direction and guidance of superiors
- no desire for improvement
- disrespectful attitude and behavior
- work marked by others, complete failure
- difficult for others to work with
- suitable for routine, ordinary jobs
- tries hard but achieves little
- leadership school was no help
- overly concerned with self-image
- far below average performer
- poor leader and follower
- leadership ability is questionable
- inattentive to details
- plagued by indecisiveness
- duty performance totally unsatisfactory
- does nothing right
- unreliable performance despite counseling
- performance far below peers
- incompetent in all areas of leadership
- careless and negligent in duty performance
- attitude and performance not in tune with pay grade
- has bad habit of being late

- has no idea how to lead
- has difficulty with team training
- planned badly for AGI
- ignores advice of superiors
- causes extra work for others
- continuing discipline problem
- indecisive under pressure
- lags behind others
- has "don't care" attitude

There you have it, a list of words to use to form your bullet comments as well as a list of bullet comments that you can add to or take from to build the comments you desire. You can even use the words and comments to write:

- letters of appreciation
- letters of recommendation (promotions)
- letters of commendation
- narrative for awards

Remember, some weak leaders may feel they have control over you because they are your rater. If you encounter such individuals, don't let them intimidate you. All you have to do is make sure you do your job the best you can and make sure you are counseled and understand what you were counseled about.

Should you receive a bad NCO-ER before you were able to do what is outlined in this book, read "Appealing the NCO-ER" in the next chapter.

5

Helpful NCO-ER Tips and Information

Army Values

It took 13 years to change the NCO-ER, which is a strong indication of its success. The NCO-ER form itself was changed from DA Form 2166-7 to DA Form 2166-8, and the counseling form was changed from DA Form 2166-7-1 to DA Form 2166-8-1. The main change of the report is in Part IVa where the new Army values are listed and the raters render their evaluation based on those values. It's very important that the raters understand the meaning of the values as well as you, the rated NCO. During the counseling session the rater should go over each of the values with the rated NCO, explaining them as well as telling what you need to do in order to get an "X" in the "YES" box of each value. Checking the "YES" box should also get you two or more good bullet comments in the box under the list of values. Bullet comments are used to explain any area where the rated NCO is particularly strong or needs improvement. Remember, the bullet comments are mandatory for a "NO" rating and must be specific. Now, let's look over the seven Army listed values.

Loyalty Means bearing true faith and allegiance to the U.S. Constitution, the Army, your unit, and other soldiers. You have an obligation to be faithful to the Army, the institution and its people, and to your unit or organization. When you started your Army career by swearing allegiance to the Constitution, the basis of our government and laws, you, in fact, were saying that you will be loyal to the United States. The loyalty of your soldiers is a gift they give to you, and you should give them that same gift. Loyalty is a commitment, it's what makes your soldiers feel good and safe about going into a combat zone with you. It's your loyalty that gives them faith that they will return home to their loved ones.

Duty Means fulfilling your obligations. Duty begins with everything required of you by law, regulation, and orders. It's taking full responsibility for your actions as well as the actions of your soldiers. It's also your duty to disobey illegal orders. If you think an order is illegal, be 100 hundred percent sure that you understood both the details of the order and the original intent. Seek clarification from the person giving the order. If you must decide immediately, make the best judgment possible based on the Army values, your experience, and your previous study and reflection. If you don't have to make an immediate decision, seek legal counsel. You take a big risk when you go alone with what is an illegal order. Disobeying an illegal order may be the most difficult decision you'll have to make in your Army career, but that's one of the things good leaders have to do.

Respect/ EO/EEO Means treating people as they should be treated. In the Army, respect means recognizing and appreciating the inherent dignity and worth of all people. Your soldiers are your greatest resource. Army leaders honor everyone's individual worth by treating all people with dignity and respect. They must be aware that they will deal with people from a wide range of ethnic, racial, and religious backgrounds. You must foster a climate in which every-one is treated with dignity and respect regardless of race, gender, creed, or religious belief. Respect is also an essential component for the development of a disciplined, cohesive, and effective war-fighting team. Respect goes beyond discrimination and harassment; it includes the broader issue of civility, the way people treat each other and those they come in contact with. Soldiers, like their leaders, treat everyone with dignity and the utmost respect.

Selfless Service

Means putting the welfare of the nation, the Army, and subordinates before your own. Selfless service means doing what's right for the nation, the Army, your unit, and your soldiers. The needs of the Army and the nation come first. This doesn't mean that you neglect your family or yourself; in fact, such neglect weakens a leader and can cause the Army more harm than good. Selfless service doesn't mean that you can't have a strong ego, high self-esteem, or even healthy ambition. Rather, self-less service means that you don't make decisions or take actions that help your image or career but sabotage the mission. A selfless NCO would take credit for the work his/her soldiers did, but the selfless NCO would give the credit to the soldiers.

Honor

Means living up to all the Army values. Implicitly, that is what you promised when you took your oath of office or enlistment. Honor provides the moral compass for character and personal conduct in the Army. The expres-sion "honorable person" refers to those whose words and deeds are beyond reproach; it refers to both the character traits an individual actually possesses and the fact that the community recognizes and respects them. Honor means demonstrating an understanding of what's right, and taking pride in the community's acknowledgment of that reputation. To be a honorable person, you must be true to your oath and maintain Army values in all you do.

Living honorably strengthens Army values, not only for yourself but for others as well. Honor for an Army leader means putting Army values above self-preservation as well; this honor s essential for creating a bond of trust among members of the Army and between the Army and the nation it serves.

Integrity   Means doing what's right legally and morally. Having integrity means being both morally complete and true to yourself. If you want to instill Army values in others, you must internalize and demonstrate them yourself. Your personal values may, and probably do, extend beyond the Army values, to include such things as political, cultural, or religious beliefs. People of integrity consistently act according to principles, not just what might work at the moment. Leaders of integrity make their principles known and consistently act in accordance with them.

Personal
Courage   Means facing fear, danger, or adversity (physical or moral). Personal courage isn't the absence of fear; rather, it's the ability to put fear aside and do what's necessary. It takes two forms, physical and moral. Physical courage means overcoming fears of bodily harm and doing your duty. Moral courage is the willingness to stand firm on your values, principles, and convictions, even when things go wrong. Moral courage is essential to living the Army values of integrity and honor every day. Moral courage often expresses itself as candor. Candor means being frank – honest, and sincere with others while keeping your words free of bias, prejudice, or malice. It also means calling things as you see them, even when you feel uncomfortable or you think it might be better for you to just keep quiet. Candor means not allowing your feelings to affect what you say about a person or situation. In combat physical and moral courage may blend together. The right thing to do may not only be unpopular, but dangerous as well.

Types of Reports

Annual

A report will be submitted 12 months after the most recent of the following events:

1. Ending month of last report.

2. Effective date of promotion to sergeant.

3. Reversion to NCO status after serving as a commissioned warrant officer for 12 months or more.

4. Reentry on active duty in a rank or sergeant or above after a break in enlisted services of 12 months or more.

The 90-day rater minimum qualification period must be met. In cases when it is not, the annual report period will be extended until the minimum rater qualification period is met. An annual report will not be signed prior to the first day of the month following the ending month of the report. An annual report will not be submitted when the provisions for the change-of-rater report apply. The senior rater will complete both the rater and senior rater portions of the NCO-ER, provided that minimum rater qualifications are met, under the following circumstances:

1. The rater dies, is relieved, reduced, or AWOL.

2. The rater is declared an unsatisfactory participant based on AR 135-91.

3. The rater is declared missing or incapacitated after the report period, but before the report is signed.

Change-of-Rater

A report will be submitted whenever the designated rater is changed, as long as minimum rater qualifications are met. The minimum rating period is 90 days. Rater changes include:

1. The rater or rated NCO is reassigned.

2. The rater or rated NCO departs on extended TDY or SD.

3. The rater or rated NCO is relieved from active duty or full-time National Guard duty early based on AR 635-200 or AR 135-178, or normal expiration term of service (ETS), except for discharge and immediate re-enlistment.

4. The rated NCO is reduced to CPL/SPC or below. Part Ic will contain the "reduced" rank, and Part Id will reflect the effective date of the reduction. Reduction to another NCO grade does not require a report, unless the actual rater changes.

5. The senior rater will complete both the rater and senior rater portions of the report, and will enter a brief explanation of the reason in Part Ve. When both the rater and senior rater are unable to render an evaluation, a report will not be submitted; code "Q" will be used to explain the non-rated period. A change-of-rater report is mandatory when the rated NCO is separated from active duty.

As an exception, retirement reports of less than one year will be rendered at the option of the rater or senior rater or when requested by the rated NCO.

The change-of-rater report may not be signed before the date the change occurs. In the event of PCS, ETS, or retirement, the report may be completed and signed up to 10 days prior to the date of departure in order to facilitate orderly out-processing.

Compassionate Reassignment, Temporary or Special Duty

When an NCO departs on temporary or special duty under one of the following conditions, a change-of-rater report for both the NCO and their eligible subordinates will be submitted, provided rater qualifications are met, prior to departure:

1. To attend a resident course of instruction or training scheduled for 90 calendar days or more at a service school.

2. To attend a civilian academic or training institution on a full-time basis for a period of 90 calendar days or more.

3. To perform duties not related to his/her primary functions in his/her parent unit under a different immediate supervisor for 90 days or more. In cases where it cannot be determined if the TDY or SD will last 90 days, a report will be submitted. A report is not authorized if the NCO will still be responsible to or be receiving instructions from a rating official in the parent organization. An NCO on TDY or SD other than 1 and 2 above who is not responsible to rating officials in his/her parent unit will be rated by the TDY or SD supervisor. The TDY or SD supervisor will ensure that a rating scheme is published. An NCO on TDY or SD who remains responsible to a rating official in the parent unit will continue to be rated for that period, regardless of its length, by the normal rating officials. Memorandum input from officials at TDY or SD locations is optional. An NCO attached to an organization pending compassionate reassignment remains responsible to his/her parent unit and will not receive an evaluation report from the attached organization. Memorandum input from the supervising officials of the attached organization is mandatory.

Relief-For-Cause

A report is required when an NCO is relieved for cause regardless of the rating period involved. Relief-for-cause is defined as removal of an NCO from a rateable assignment based on decision by members of the NCO's chain of command or supervisory chain. A relief-for-cause occurs when the NCO's personal or professional characteristics, conduct, behavior, or performance of duty warrants removal in the best interest of the U.S. Army. If for some reason the relief does not occur on the date the NCO is removed from his/her duty position or responsibilities, the suspended period of the time between the removal and the relief will be non-rated time included in the period of the relief report. The published rating chain at the time of the relief will render the report. No other report will be due during this non-rated period. When the NCO is suspended from duties pending investigation, every effort should be made to retain the established rating chain until the investigation is resolved. If relief-for cause is contemplated on the basis of an informal AR 15-6 investigation, the referral procedures contained in that regulation must be complied with before the act of initiating or directing the relief. A relief-for-cause should be the final action after all investigations have been completed and a determination made.

Rules for the Relief-For-Cause Report

1.   The rating official directing the relief will clearly explain the reason for the relief in Part IV, if the relieving official is the rater of Part Ve or if the relieving official is the senior rater.

2.   If the relief is directed by an official other than the rater or senior rater, the official directing the relief will describe the reason for the relief in an enclosure to the report.

3.   Regardless of who directs the relief, the rater will enter the bullet, "the rated NCO has been notified of the reason for the relief" in Part IVf.

4.   The minimum rater and senior rater qualifications and minimum rating period are 30 rated days. This restriction allows the rated NCO a sufficient period to react to performance counseling during the rating period. Authority to waive this 30-day minimum rating period and rater and senior rater qualification period in cases of misconduct is granted to a general with court-martial jurisdiction over the relieved NCO. The waiver approval will be in memorandum format and attached as an enclosure to the report.

5.   The date of relief determines the "THRU" date of the report. Relief-for-cause reports may be signed at anytime during the closing or following month of the report.

6. When the rater is relieved, or when the rated NCO and the rater are concurrently relieved, the senior rater will complete the rater and senior rater portions of the report for each of the rater's subordinates. Enter "rater relieved" in Part Ve, and do not identify the relieved rater in Part Iia.

7. DA Form 2166-8, Part I, Item j will reflect one rated month when the computation of rated months results in zero. Cases where the rated NCO has been suspended from duties pending an investigation should be resolved by the chain of command as expeditiously as possible to reduce the amount of non-rated time.

Complete-the-Record Report

At the option of the rater, a complete-the-record report may be submitted on an NCO who is about to be considered by a DA centralized board for promotion, school, or CSM selection, provided the following conditions are met:

1. The rated NCO must be in the zone of consideration (primary or secondary) for a centralized promotion board or in the zone of consideration for a school or CSM selection board.

2. The rated NCO must have been under the same rater for at least 90 rated days as of the ending month established in the message announcing the zone of consideration.

3. The rater must not have received a previous report for the current duty position.

Complete-the-record reports will not be signed prior to the first day of the month following the ending month. Complete-the-record reports are optional. Therefore, the absence of such a report from the Official Military Personnel File (OMPF) at the time of the board's review will not be a basis for request standby reconsideration unless the absence is due to administrative error or a delay in processing at the Enlisted Evaluation Center (EREC).

Senior Rater Option

When a change in senior rater occurs, the senior rater may direct that a report be made on any NCO for whom he/she is the senior rater, but only if the following conditions are met:

1. The senior rater has served in that position for at least 60 rated days. In cases where a general officer is serving as both rater and senior rater, the minimum rater requirement will also be 60 rated days versus the normal 90 days requirement.

2. The rater meets the minimum requirement to give a report.

3. The rated NCO has not received a report in the preceding 90 rated days.

If an evaluation report would become due within 60 calendar days after the change in senior rater, the senior rater will submit a senior rater option report to prevent an NCO-ER being submitted later without a senior rater evaluation, provided the above requirements (1, 2, 3) are met.

Sixty-Day Option

A report may be submitted at the option of the rater if there are fewer than 90 rated days but more than 59 rated days in the rated period, and the following conditions are met:

1. The rated NCO must be serving in an overseas designated short tour for a period of 14 months or less.

2. The senior rater must meet the minimum time-in-position requirement to evaluate (60 rated days) and must approve or disapprove submission of the report. When the senior rater disapproves the submission of the report, he/she will state the basis for the disapproval and return the report to the rater, at which time the rater will inform the rated NCO that the report has been disapproved and then he/she (rater) will destroy the report.

The Counseling Process

The counseling process consists of four stages:

1. Identify the need for counseling
2. Prepare for counseling
3. Conduct counseling
4. Follow up

Quite often organizational policies, associated with an evaluation required by the command, demand a counseling session. Developing soldiers consists of observing the soldiers' performance, comparing it to the standards, and then providing feedback to the soldiers in the form of counseling. To prepare for counseling the leader must:

- Select a suitable place
- Schedule the time
- Notify the soldier well in advance
- Organize information
- Outline the counseling session components
- Plan your counseling strategy
- Establish the right atmosphere

Schedule counseling in an environment that minimizes interruptions and is free from distraction, sights, and sounds. Schedule the time so that the counseling is done during duty time. The session should last less than an hour; if more time is needed, schedule a second session. Select a time that is free from conflict with other activities.

The soldier should know why, where, and when the counseling will take place. Counseling should happen as close to the event as possible. Solid preparation is also essential to effective counseling, which includes the purpose, facts and observations about the soldier, identification of possible problems, main points of discussion, and the development of a plan of action. Using the information obtained, determine what to discuss during the counseling session:

- Note what prompted the counseling
- Know what you aim to achieve

- Know what your role as a counselor is
- Identify possible comments or questions for the counseling session.

A written outline will help organize the session and promote positive results. The right atmosphere promotes two-way communication between a leader and soldier. You may offer the soldier a seat or a cup of coffee. Sit in a chair facing the soldier since a desk can become a barrier. During counseling to correct substandard performance, you may direct the soldier to remain standing while you remain seated behind the desk. Giving specific guidance (formal atmosphere) reinforces the leader's rank, position in the chain of command, and authority.

Conducting the Counseling Session

You must be flexible when conducting the counseling session. Often counseling for a specific incident occurs as the leader encounters the soldiers in their daily activities. Such counseling can occur in the fields, motor pool, barracks, or wherever one performs duties. You should always address the four basic components of a counseling session which consist of the following:

- Opening the counseling session
- Discussing the issues
- Developing the plan of action
- Recording and closing the session

Opening the Counseling Session

The best way to open the counseling session is to clearly state its purpose. Establish a soldier-centered setting by inviting the soldier to speak. If applicable, start the counseling session by reviewing the status of the previous plan of action. Attempt to develop a mutual understanding of the issue by letting the soldier do most of the talking. Aim to help the soldier better understand the subject of the counseling. When the issue is standard performance, explain how his/her performance didn't meet the standards.

Developing the Plan of Action

A plan of action pinpoints desired results. It specifies what the soldier must do to reach the goals set during the counseling session. The plan of action should show the soldier how to modify or maintain his/her behavior. It must use concrete and direct terms and set the stage for successful development.

Recording and Closing the Session

Documentation serves as a reference to the agreed-upon plan of action, as well as the soldier's accomplishments, improvements, personal preferences, or problems. Army regulations require written records of counseling for certain personnel actions, such as:

- Bar to re-enlistment
- Processing a soldier for administrative separation
- Overweight program
- Substandard duty performance

Substandard Duty Performance

Documentation conveys a strong corrective message to soldiers. To close the session, summarize its key points and ask if the soldier understands the plan of action. Establish any follow-up measures to support the plan of action. Schedule any future meeting, at least tentatively, before dismissing the soldier.

Leader's Follow-up Responsibilities

Closing the session is not the end of the counseling process, which continues through implementation of the plan of action and evaluation of results. You must support soldiers as they implement their plan of action. Support may include teaching, coaching, or providing time and resources. You must observe and assess this process and possibly modify the plan of action to meet its goals. Appropriate measures after counseling include:

- Follow-up counseling
- Making referrals
- Informing the chain of command
- Taking corrective measures

Remember, the purpose of counseling is to develop soldiers who are better able to achieve personal, professional, and organizational goals.

Developmental Counseling

Developmental counseling is soldier-centered communication that produces a plan outlining actions that soldiers must take to achieve individual and organizational goals. Soldier leadership development is one of the most important responsibilities of every Army leader. Just as training includes AARs (After Action Reports) to fix shortcomings, leadership development includes performance reviews, which result in agreement between the leader and soldier on a development strategy or plan of action that builds on the soldier's strengths and establishes goals to improve on weaknesses. Developmental counseling is a shared effort. As a leader you should assist your soldiers in identifying strengths and weaknesses and creating plans of action. To achieve success, your soldiers must be forthright in their commitment to improve and be candid in their assessment and goal setting.

You as a Counselor

You must demonstrate certain qualities to become an effective counselor, such as respect for your soldiers, self-awareness and cultural awareness, empathy, and credibility. You show respect for your soldiers when you allow them to take responsibility for their actions. You must be fully aware of your own values, needs, and biases prior to counseling your soldiers. Cultural awareness is a mental attribute, that enhances your ability to display empathy. You must be aware of the similarities and differences between individuals of different cultural backgrounds and how these factors could influence values, perspectives, and actions. Empathy is the action of being understanding of, and sensitive to, the feelings, thoughts, and experience of another person to the point that you can almost feel or experiences them yourself. By understanding the soldiers' position you can help them develop a plan of action that fits their personality and needs. If you don't fully comprehend a situation from your soldiers' points of view, you have less credibility and influence and your soldiers are less likely to commit to the agreed- upon plan of action. Leaders achieve credibility by being honest and consistent in their statements and actions. If you lack credibility with your soldiers' you'll find it difficult to influence them. One way to earn credibility is be consistent in what you do and say.

Counselor Counseling Skills

The techniques you use for effective counseling must fit the situation. You may only need to provide information, listen, or give words of praise. Three of the best ways to improve your counseling techniques include:

- Studying human behavior
- Learning the kinds of problems that affect your soldiers
- Developing your interpersonal skills

General skills that you'll need in almost every situation include active listening, responding, and questioning. When you've actively listened while counseling your soldiers, you've received their message. Some elements of active listening are:

- **Eye contact.** Maintaining eye contact without staring helps show sincere interest. The occasional breaks of eye contact are normal and acceptable.

- **Body posture.** Being relaxed and comfortable will help put the soldier at ease, but a too-relaxed position or slouching may be interpreted as a lack of interest.

- **Head nods.** Nodding your head shows you're paying attention and encourages the soldier to continue.

- **Facial expressions**. A blank look or fixed expression may disturb the soldier, whereas smiling too much or frowning may discourage him/her from continuing.

- **Verbal expressions**. Let your soldiers do the talking, but keep the discussion on the counseling subject. A long silence can sometimes be distracting and make the soldier feel uncomfortable.

Active listening means listening thoughtfully and deliberately to the way a soldier says things. An opening and closing statement as well as recurring references may indicate the soldier's priorities. Pay attention to gestures, and you will "see" the feeling behind the words. Note the differences between what is said and done.

Nonverbal indicators of the future leader's attitude includes:

- **Boredom**. Drumming on the table, doodling, clicking a ballpoint pin, or resting the head in the palm of the hand.

- **Self-confidence**. Standing tall, leaning back with hands behind the head, and maintaining steady eye contact.

- **Defensiveness**. Pushing deeply into a chair, glaring at the leader, and making sarcastic comments as well as crossing or folding arms in front of the chest.

- **Frustration**. Rubbing eyes, pulling on an ear, taking short breaths, and frequently changing body position.

- **Interest, friendliness, and openness**. Leaning toward the leader while sitting.

- **Anxiety**. Sitting on the edge of the chair with arms uncrossed and hands open.

Consider these indications carefully. Each may show something about the soldier, but don't assume – ask the soldier about the indicator so you can better understand his/her behavior. Responding skills follow-up on active skills. From time to time check your understanding; clarify and confirm what has been said. Respond both verbally and nonverbally. Verbal responses consist of summarizing, interpreting, and clarifying while nonverbal responses include eye contact, facial expressions, and occasional gestures such as a head nod. Questioning is a necessary skill, which you must use with caution. Too many questions may place the soldier in a passive mode. Ask questions to obtain information or to get the soldier to think about a particular situation. Well-posed questions may help to verify understanding, encourage further explanation, or help the soldier move through the stages of the counseling session.

Counseling Errors

You should never dominate the counseling by talking too much, giving unnecessary inappropriate "advice," not truly listening, and projecting personal likes, biases, and prejudices, all which interferes with effective counseling. You should also avoid rash judgment, stereotypes, loss of emotional control, inflexible methods of counseling, and improper follow-up. To improve your counseling skills you should:

- Determine the soldier's role in the situation and what the soldier has done to resolve the problem or improve performance .

- Draw conclusions based on more than the soldier's statement

- Try to understand what the soldier says and feels; listen to what the soldier says and how he/she says it.

- Show empathy when discussing the problem.

- Ask questions, questions that are relevant to the discussion.

- Keep the conversation open-ended; avoid interruption.

- Give the soldier your full attention.

- Be receptive to the soldier's feelings but you should not try to save the soldier from his/her emotional pain.

- Encourage the soldier to take the initiative and to say what is on his/her mind.

- Avoid an interrogating manner; listen more and talk less.

- Keep your personal experiences out of the counseling session unless you believe sharing your experiences will be helpful.

- Avoid confirming a soldier's prejudices; remain objective.

- Help the soldier help himself/herself.

- Know what information to keep confidential and what to present to the chain of command.

Appealing the NCO-ER

A bad NCO-ER is one that has one or more "Needs Improvement" bullet comments, including a check in the "NO" box of #3 (Respect/EO/EEO) in Part IV of DA Form 2166-8. Should you get a report with a "Need Improvement" comment, you should first talk to the rater who gave it to you to see if he/she will change it. If that does not work, talk to the reviewer, or whomever is next in the rating chain; do this only if you feel you should have received a better bullet comment. The next step you may want to take is:

- Talk to the personnel CSM about the report.

- Talk to the local staff judge advocate about it.

- Talk to the NCO in charge of NCO-ERs at the PSB.

To determine whether an appeal is advisable, the PSB personnel can also provide assistance in the preparation of the appeal process if need be. Be realistic as to whether or not you want to submit an appeal; some NCOs may not perform certain duties as well as others. Appealing an NCO-ER just because you disagree will most likely be unsuccessful. Also, careful consideration should be given before appealing an NCO-ER in which the narrative portions are good but the numerical marking or box checks are less than maximum. You can seek an initial means of redress through a Commander's Inquiry, however this is not a prerequisite for submitting an appeal.

Commander's Inquiry

The primary purpose of the Commander's Inquiry is to provide a greater degree of command involvement in preventing obvious injustices of the rated NCO and correct errors before they become a matter of permanent record. If you want to have a Commander's Inquiry prepared, the commander should appoint an investigating officer to conduct the inquiry and report the finding/recommendations BEFORE the NCO-ER is processed. Once the inquiry is completed, a completed copy of the inquiry, along with all attachments and the commander's finding, and the original NCO-ER, should be forwarded to DA for the final review and processing. The address is:

> CDR DA PERSCOM
> ATTN: TAPC-MSE
> 200 Stovall Street
> Alexandria, VA 22332-0400

The Commander's Inquiry will not be used to document differences of opinion among rating officials or between the commander and rating officials about an NCO's performance and potential. Rating officials should evaluate and have their opinions constitute the organization's view of the rated NCO; however, the commander may determine through the inquiry that the report has serious irregularities or errors. Some examples include, but are not limited to:

- Improperly designated or unqualified rating officials
- Inaccurate or untrue statements
- Lack of objectivity or fairness by rating officials

The Commander's Inquiry will be made by a commander (Major or above) in the chain of command above the rating official involved in the allegations. The commander will confine the inquiry to matters related to:

- The clarity of the report
- The facts contained in the report
- The compliance of the report IAW AR 623-205
- The conduct of the rated NCO and rating officials.

The procedures for the inquiry may be as formal or informal as the commander deems appropriate, to include telephone and personal discussions. The commander may also appoint an officer senior to the designed rating officials involved in the allegation. To make the inquiry, the commander:

- Will not force or pressure the rating official to change his/her evaluation.

- May not evaluate the rated NCO, either as a substitute for, or in addition to, the designed rating official's evaluation.

- Will not use the inquiry provisions to forward information derogatory to the rated NCO.

The inquiry must be conducted by either the commander who is still in the command position at the time the report was rendered or by a subsequent commander in the position. The results of a Commander's Inquiry will include findings, conclusions, and recommendations in a format that can be filed with the report in the NCO's personnel file (OMPF) for clarification purpose. The results, therefore, will include the commander's signature, should stand alone without reference to the other documentation, and will be limited to one page. Sufficient documentation, such as reports and statements, will be attached to justify the conclusion. Remember the results of a Commander's Inquiry do not constitute an appeal, but can be used to support an appeal.

Preparing an Appeal

An appeal's success depends on the care with which the case is prepared, the line of argument presented, and the strength of the evidence presented to support it. Start by identifying those comments to be challenged, the perceived inaccuracy in each entry or comment, the evidence you think is necessary to prove the alleged inaccuracy, and where and how to obtain such evidence. You should carefully decide what evidence is needed to support claims, whether or not such evidence is available, and how to go about obtaining it. After considering the nature of a claim, and you still believe the evaluation report is inaccurate and evidence is available to support your argument, you should prepare and submit an appeal.

Obtaining Evidence

Collect supporting evidence necessary to adequately refute and contest the evaluation report. Third parties are persons who have official knowledge of the rated NCO's duty performance during the period of the report being appealed. Third-party statements form the basis of most substantive appeals. Statements from individuals who establish they were on hand during the contested rating period, who refute fault finding remarks on the evaluation report, and who served in position from which they could observe the NCO's performance and his/her interactions with the rating official, are both useful and supportive. These statements should be specific and not deal in general discussions of the NCO. Although third-party statements can be provided by knowledgeable subordinates, peers, and superiors, additional weight is normal given those statements where the authors occupied vantage points during the contested period that closely approximated those of the rating officials. Such third-party statements should be on letterhead if possible, describing the author's duty relationship to the NCO during the period of the contested report, degree (frequency) of observation, and should include the author's current address and telephone number. Statements from rating officials often reflect retrospective thinking, or second thoughts, prompted by an NCO's non-selection or other unfavorable personnel action claimed to be the sole result of the report. As a result, claims by rating officials that they did not intend to evaluate as they did will not, alone, serve as the basis of alerting or withdrawing an evaluation report. Official documents may substantiate that an evaluation report is in error. Such documents include but are not limited to:

- An administrative appeal
- A certified copy of a published rating scheme
- Duty appointment orders
- Appropriate extracts from local personnel records
- Annual General Inspection results
- Award citation and letters of commendation

To obtain current mailing addresses of Army personnel, check with your local PSB to see if your installation has a copy of the U.S. Army locator for members on active duty or state personnel roster. If this is not available, call the worldwide locator service [(DSN) 221-3732] or commercial (703-325-3732) which is open 24/7, or you can write to the active Army locator. Include the full name and Social Security Number (SSN) of those you wish to contact. The address is:

> Army World Wide Locator
> 8899 East 56th Street
> Indianapolis, IN 46249-5301

If the person you are trying to contact is not on file or has retired from the Army or left active duty, write to:

> National Personnel Records Center
> 9700 Page Avenue
> St. Louis, MO 63132-5260

The person's full name and SSN must be provided along with the request. State that this is for official use, that is – in conjunction with an NCO-ER appeal. Relevant portions of official documents such as Annual General Inspection (AGI), Army Training and Evaluation Program (ARTEP), or command inspection results may be obtained under the Freedom of Information Act (FOIA) by writing the individual unit or headquarters responsible for conducting such inspections. Addresses for military organizations can be obtained by contacting your service PSB. To obtain records to verify dates, start with your 210 file for orders and other, or contact the former organization PSBs or unit level personnel office to determine whether records are still retained.

Cover Memorandum and Appeal Format

The appeal memorandum should be typed in memorandum format on letterhead or written on bond paper. Identify in the first paragraph the name, PMOS, SSN, period of record and priority of appeal. Include a DSN or commercial phone number and correct mailing address. A home address may be used, if preferred. Use this memorandum as the basis of the appeal. Identify the specific portions of the report being contested. Be clear, brief, and specific. If detailed information is essential, add a statement as an enclosure to the appeal. Indicate the specific change requested – that is, a single change, a combination of changes, or total removal of the report. All enclosures should be tabbed and listed for reference purposes and cited in the written presentation of the case. Sign and date the cover memorandum.

Submission

Before finalizing the appeal, the NCO should have the entire package reviewed by a disinterested party in whom he/she has trust and confidence. This third-party review may help remove emotionalism and poor use of logic from the case. Don't forget your PSB is there to help you with all the paper work. The appeal package should not be submitted until you are satisfied that you have put together a logical, well-constructed case, and as fully documented as possible. Submit the finalized appeal in two complete packets to:

> CDR U.S. Army Enlisted Records
>   and Evaluation Center
> ATTN: PCRE-REA
> 8899 East 56th Street
> Indianapolis, IN 46249-5301

Career Resources

THE FOLLOWING CAREER RESOURCES are available directly from Impact Publications. Full descriptions of each title as well as nine downloadable catalogs, videos, and software can be found on our website: www.impactpublications.com. Complete the following form or list the titles, include shipping (see formula at the end), enclose payment, and send your order to:

IMPACT PUBLICATIONS
9104 Manassas Drive, Suite N
Manassas Park, VA 20111-5211 USA
1-800-361-1055 (orders only)
Tel. 703-361-7300 or Fax 703-335-9486
Email address: info@impactpublications.com
Quick & easy online ordering: www.impactpublications.com

Orders from individuals must be prepaid by check, money order, or major credit card. We accept telephone, fax, and e-mail orders.

Qty.	TITLES	Price	TOTAL

Books by Author

____	Becoming a Better Leader and Getting Promoted in Today's Army	$13.95	____
____	Complete Guide to the New NCO-ER	13.95	____
____	The NCO-ER Leadership Guide	14.95	____
____	Up or Out: How to Get Promoted As the Army Draws Down	13.95	____

Military- and Government-Related Resources

____	The Book of U.S. Government Jobs	21.95	____
____	The Complete Guide to Public Employment	19.95	____

_____	Directory of Federal Jobs and Employers	21.95	_____
_____	Electronic Federal Resume Guide	44.95	_____
_____	FBI Careers	18.95	_____
_____	Federal Application Kit	35.95	_____
_____	Federal Applications That Get Results	23.95	_____
_____	Federal Employment From A to Z	14.50	_____
_____	Federal Resume Guidebook	21.95	_____
_____	Find a Federal Job Fast!	13.95	_____
_____	From Army Green to Corporate Gray	17.95	_____
_____	Jobs and the Military Spouse	17.95	_____
_____	Ten Steps to a Federal Job	39.95	_____

SUBTOTAL   _____

Virginia residents add 4½% sales tax   _____

POSTAGE/HANDLING ($5 for first
product and 8% of SUBTOTAL) $5.00

8% of SUBTOTAL ----------------------------- _____

TOTAL ENCLOSED --------------- _____

SHIP TO:

NAME _____

ADDRESS _____

_____

_____

PAYMENT METHOD:

❑ I enclose check/money order for $ _____ made payable to
IMPACT PUBLICATIONS.

❑ Please charge $ _____ to my credit card:
❑ Visa ❑ MasterCard ❑ American Express ❑ Discover

Card # _____ Expiration date: _____/_____

Signature _____

Keep in Touch . . .
On the Web!

www.impactpublications.com
www.ishoparoundtheworld.com
www.travel-smarter.com
www.contentfortravel.com
www.winningthejob.com
www.veteransworld.com
www.contentforcareers.com